GOOD FOOD VS. BAD FOOD

Other Books by the Author

THE BIKINI DIET

20/20 THINKING

FOODS THAT COMBAT CANCER

CONTROL DIABETES IN SIX EASY STEPS

WRINKLE-FREE: YOUR GUIDE TO YOUTHFUL SKIN AT ANY AGE

THE BONE DENSITY TEST

THE CELLULITE BREAKTHROUGH

HAIR SAVERS FOR WOMEN:
A COMPLETE GUIDE TO TREATING AND PREVENTING HAIR LOSS

NATURAL WEIGHT LOSS MIRACLES

21 DAYS TO BETTER FITNESS

KAVA: THE ULTIMATE GUIDE TO NATURE'S ANTI-STRESS HERB

Other Books Coauthored by the Author

LEAN BODIES

LEAN BODIES TOTAL FITNESS

30 DAYS TO SWIMSUIT LEAN

HIGH PERFORMANCE NUTRITION

POWER EATING

SHAPE TRAINING

HIGH PERFORMANCE BODYBUILDING

50 WORKOUT SECRETS

BUILT! THE NEW BODYBUILDING FOR EVERYONE

GOOD FOOD
vs. BAD FOOD

Maggie Greenwood-Robinson, Ph.D.

BERKLEY BOOKS, NEW YORK

THE BERKLEY PUBLISHING GROUP
Published by the Penguin Group
Penguin Group (USA) Inc.
375 Hudson Street, New York, New York 10014, USA
Penguin Group (Canada), 90 Eglinton Avenue East, Suite 700, Toronto, Ontario M4P 2Y3, Canada
(a division of Pearson Penguin Canada Inc.)
Penguin Books Ltd., 80 Strand, London WC2R 0RL, England
Penguin Group Ireland, 25 St. Stephen's Green, Dublin 2, Ireland (a division of Penguin Books Ltd.)
Penguin Group (Australia), 250 Camberwell Road, Camberwell, Victoria 3124, Australia
(a division of Pearson Australia Group Pty. Ltd.)
Penguin Books India Pvt. Ltd., 11 Community Centre, Panchsheel Park, New Delhi—110 017, India
Penguin Group (NZ), Cnr. Airborne and Rosedale Roads, Albany, Auckland 1310, New Zealand
(a division of Pearson New Zealand Ltd.)
Penguin Books (South Africa) (Pty.) Ltd., 24 Sturdee Avenue, Rosebank, Johannesburg 2196,
South Africa

Penguin Books Ltd., Registered Offices: 80 Strand, London WC2R 0RL, England

Every effort has been made to ensure that the information contained in this book is complete and accurate. However, neither the publisher nor the author is engaged in rendering professional advice or services to the individual reader. The ideas, procedures, and suggestions contained in this book are not intended as a substitute for consulting with your physician. All matters regarding your health require medical supervision. Neither the author nor the publisher shall be liable or responsible for any loss or damage allegedly arising from any information or suggestion in this book. The recipes contained in this book are to be followed exactly as written. The publisher is not responsible for your specific health or allergy needs that may require medical supervision. The publisher is not responsible for any adverse reactions to the recipes contained in this book. The publisher does not have any control over and does not assume any responsibility for author or third-party websites or their content.

PRINTING HISTORY
Good Fat vs. Bad Fat Berkley mass-market edition / January 2003
Good Carbs vs. Bad Carbs Berkley mass-market edition / January 2004
Two-in-one Berkley trade paperback edition / January 2007

Library of Congress Cataloging-in-Publication Data

Greenwood-Robinson, Maggie.
    [Good fat vs. bad fat]
    Good food vs. bad food / Maggie Greenwood-Robinson.—Two-in-one Berkley trade pbk. ed.
    p. cm.
    First work originally published: Good fat vs. bad fat. New York, N.Y.: Berkley Books, 2002; 2nd work originally published: Good carbs vs. bad carbs. New York, N.Y.: Berkley Books, 2004. With slight corrections.
    Includes bibliographical references (p. 353)
    ISBN-13: 978-0-425-21359-9
    1. Nutrition. 2. Lipids in human nutrition. 3. Carbohydrates in human nutrition. I. Greenwood-Robinson, Maggie. Good carbs vs. bad carbs. II. Title. III. Title: Good food vs. bad food.

RA784.G758 2007
613.2'84—dc22

2006049234

PRINTED IN THE UNITED STATES OF AMERICA

10 9 8 7 6 5 4 3 2

• CONTENTS •

GOOD CARBS
vs.
BAD CARBS

To Dad, with love

• ACKNOWLEDGMENTS •

Good Carbs vs. Bad Carbs has become a reality due to the effort and energy of the following people: Christine Zika and the staff at Berkley Books; Madeleine Morel, my literary agent; and my husband, Jeff, who has given me the encouragement and support to persevere in my career. To these people, I am very grateful.

I would also like to thank a number of fine organizations that gave me permission to reprint their healthy, delicious carbohydrate recipes: the American Association of Cancer Research, the California Artichoke Advisory Board, the California Avocado Commission, the California Kiwifruit Commission, the Northarvest Bean Growers Association, the Pioneer Valley Growers Association, the U.S. Highbush Blueberry Council, and the U.S. Potato Board. I have included their websites in chapter 12 so you can try some of their many other fabulous recipes.

Part I

Getting to Know
Your Carbohydrates

· O N E ·

The Food Fuel

A big battle has been raging on the diet and nutrition front—the battle between "good" carbohydrates and "bad" carbohydrates. The point of contention has to do with this: Not all carbohydrates are created equal. For instance, there are some that will help prevent all sorts of life-shortening illnesses, from obesity to heart disease to cancer. Still others (like sugar and junk food) can set into motion harmful biochemical reactions inside your body that incite disease.

Obviously, if you want to stay healthy and in terrific shape for as long as you can, you should eat more carbohydrates from the first group and less from the second. And that's exactly what this book helps you do. It helps you discern between foods containing good carbohydrates (natural foods and fiber), which confer enormous health benefits, and bad carbohydrates (sugar, processed foods, and junk carbs), which are less valuable and, in excess quantities or unbalanced dietary proportions, can actually hurt your health and lead to unsightly weight gain.

When you begin to understand the differences among various types of carbohydrates, you can make better food choices and enhance your health, lose weight, boost your physical and mental performance, and protect yourself against disease or its progression. As you read through these pages, be prepared for a few startling surprises about carbs:

- How to boost your brainpower, feel upbeat practically all the time, and expand your power of recall—all with good carbs.

- A way for you to lose weight permanently and *not* cut carbs from your diet. In fact, there are certain carbs you must eat if you want to get on the path to lifelong slimness and good health.

- Scientific proof that sugar might be more damaging to your heart than even fat.

- Ways to feel fully energized to give your body the fat-burning boost it needs so you look and feel younger.

- How to plan your meals with a variety of certain types of carbs—and live a lot longer.

If you are ready to start living with more control over your weight and your health than ever before, the information in this book will get you to your goals. With this in mind, let's get started with a short nutritional lesson on how carbohydrates, in general, work in your body.

Carbohydrates 101

Whether it is the cereal you had for breakfast or the roll you ate at dinner, all carbohydrates are organic compounds containing carbon, hydrogen, and oxygen. Carbohydrates are formed directly or indirectly from green plants through a process called "photosynthesis," which you probably learned about in grade school. In this process, the sun's energy is captured by chlorophyll, the green coloring in leaves, to turn water from the soil and carbon dioxide from the air into an energy-yielding sugar called glucose. Plants require glucose for growth and repair, and any extra, unneeded glucose is converted to starch and stored in the plant.

Carbohydrates are amazingly diverse, present in foods not only as sugars and starch, but also as fiber. Sugars are sometimes called "simple" sugars, a term that describes their chemical structure. Simple sugars, for example, are constructed of either single (monosaccharide) or double (disaccharide) molecules of sugar. The three major single sugars are glucose, fructose, and galactose. Glucose and fructose are found mostly in fruits and vegetables, and galactose is a component of milk and other dairy products.

When two single sugars link up chemically, a double sugar is formed. The most common double sugar is sucrose, better known as table sugar. Sucrose is composed of glucose and fructose and is one of the sweetest of all sugars. Found in sugar cane and sugar beets, sucrose is purified and refined to provide the various sugar products you see on grocery store shelves: table sugar, candy, cakes, cookies, and other sweets.

A pairing of galactose and glucose yields the milk sugar lactose, found in dairy foods. Another double sugar is maltose, which is constructed from two units of glucose. Maltose is found in plants during the early stages of germination. Sugars, regardless of their molecular makeup, are very easily and quickly digested in the body.

Starches are created when three or more glucose molecules hook up. Most plant foods, including cereals, whole grains, pasta, fruits, and vegetables, are starches, also known as complex carbohydrates. Starches take longer to digest than sugars do.

Found in starches and in less-starchy vegetables like lettuce, fiber is the indigestible remnant of plant food. Although it contributes little energy to your diet, fiber provides bulk, which is vital for intestinal action. Fiber has an array of health benefits: It improves elimination, flushes cancer-causing substances from the system, and helps normalize cholesterol levels. A diet high in fiber will help you control your weight, too—in several ways. Fiber makes you feel full so you don't overeat. More energy (calories) is spent digesting and absorbing high-fiber foods. In fact, if you increase your fiber intake to 35 grams a day, you'll automatically burn 250 calories a day, without exercising more or eating less. And fiber helps move food through your system more efficiently. This means fewer calories are left to be stored as fat.

Making up roughly half the calories consumed in the average American diet, carbohydrates are to your body what gas is to your car—the fuel that gets you going. During digestion, all sugars and starches are broken down into glucose (blood sugar) for energy. Assisted by a hormone called insulin, blood glucose is then ushered into cells to be used by various tissues in the body. Carbohydrates, therefore, nourish your body's tissues, providing energy for your brain, central nervous system, and muscle cells in the form of glucose.

Several things can then happen to glucose in your body. Once inside a cell, it can be quickly metabolized to supply energy. Or it might be converted to either liver or muscle glycogen, the storage form of carbohydrate. When you exercise or use your muscles, your body mobilizes muscle glycogen for energy. Blood glucose can also turn into body fat and get packed away in fat tissue. This happens when you eat more carbohydrates than you need or than your body can store as glycogen. Some blood glucose might also be excreted in your urine.

Above all other nutrients, your body prefers to run on glucose from carbohydrates, even though two other nutrients, fat and protein, can provide energy, too. Fat is considered a food fuel, but it is more of an understudy or backup source. Only if carbohydrate stores dwindle will your body tap in to fat for fuel.

Protein is not a good source of energy for your body and is, in fact, underemployed when called upon as an energy provider. Its most important job in your body is to build and repair tissue. Using protein as an energy source is like hiring a computer specialist, then relegating him to the mailroom. Protein simply has more important jobs to do in the body than supply energy.

If you don't eat enough carbohydrates or fat, however, protein will be used as an energy source, because energy production takes metabolic priority over tissue-building. But the problem is that food protein and body protein (muscle and other tissue) will be sacrificed to supply energy, causing parts of your body to waste away. So you see, carbs hold fort as your body's best supply of energy.

Meet the Carbohydrate Foods

Getting familiar with the foods that supply the carbohydrates your body needs is the first essential step toward making healthier food choices. Here is an overview of the main carbohydrate food categories.

Grains and Cereals

Since ancient times, grains and cereals have been considered vital to life, signifying their remarkable nutritional value. They are "near-complete" foods because they contain protein, carbohydrates, and beneficial fats.

But the benefits don't stop there. Grains are loaded with a slew of trace minerals, including blood-building iron. Grains keep cholesterol levels in check; exert a cancer-protective effect on the body, particularly on the digestive system; contain blood-pressure-lowering nutrients such as potassium; and are packed with B vitamins, which are involved in metabolism (your body's food-to-fuel process).

Grains and cereals come from grasses that yield edible seeds. The most commonly cultivated grains are wheat, rice, rye, oats, barley, corn, and sorghum. Generally, these are available in their raw grain form or as ingredients in other foods such as breads and prepared cereals.

One of the oldest cultivated plants is wheat, the most widely grown grain today. Wheat is used to produce breakfast cereals and flour for various types of bakery products and pastas. In its unprocessed form, wheat is high in B vitamins and fiber. However, the milling process that refines wheat into various commercial forms removes the wheat bran and wheat germ, where the beneficial nutrients are most plentiful.

The second largest grain crop in the world is rice, a staple of all Asian countries. Unlike wheat, which is grown on large farms and cultivated mechanically, rice is grown on small paddies (fields that are submerged in water) and harvested by hand. Rice is a popular food on its own and available in many varieties. It is also used in breakfast cereals and to make alcoholic beverages such as sake. Rice that has been processed to remove only the husks is known as brown rice. It is rich in calcium, iron, and several B vitamins. White rice has been milled to remove the bran and, thus, contains fewer nutrients.

Originating in southwestern Asia around 6500 B.C.E., rye is used mostly in bread and other bakery products, as well as in distilled liquors. Nutritionally, rye supplies small amounts of potassium and B vitamins. Most of the world's rye is now grown in Poland.

A very popular grain is oat, widely available in breakfast cereals and some breads. The oatmeal you eat for breakfast is actually the flattened kernel of the oat seed with the hulls removed. Another derivative of the grain, oat flour, is used in cookies and puddings. Highly nutritious, oat grains also contain protein and fat. They are an excellent source of calcium and iron, as well as various B vitamins.

Barley is among the most ancient of grains, dating back to 5000 B.C.E. in Egypt. It was the chief grain of the Hebrews, Greeks, and Romans through the sixteenth century. Among the most nutritious forms of barley is pearled barley, made of whole kernels from which the outer husk and part of the bran have been removed. Barley supplies the malt used in the brewing of beer and the distillation of liquor. It is also an ingredient in various breakfast foods. Barley is high in calcium and phosphorus.

Originally cultivated by Indians in the Americas, corn was later introduced to Europe by Columbus and other explorers. There are many different varieties of corn, including yellow, white, red, blue, pink, and black. Corn is eaten as a fresh, canned, or frozen food and is widely used in other foods, including popcorn, tortillas, and various bakery products. In nutritional value, however, corn is inferior to other grains, lacking in protein and the B vitamin niacin.

Although principally an animal feed, sorghum is used to make a sweet syrup for cooking and baking. In many countries around the world, it is ground into a meal that can be made into porridge, flatbreads, and cakes.

There are also various specialty grains that have gained favor over the years for their high nutritional value. One of these is quinoa (*KEEN-wa*), which is technically an herb, although it looks like a grain. Quinoa, available in breakfast cereals, is unusual in that it is supremely high in protein. Unlike most grains, it contains all nine essential amino acids, the vital protein components of food.

Other specialty grains include kasha (buckwheat groats), which can be used as a side dish or a meat substitute; and couscous, a popular Mideastern grain.

Starchy Vegetables

Foods in this category include potatoes, sweet potatoes, yams, beets, and winter squash. These foods are brimming with vitamins, minerals, fiber, and health-building plant chemicals.

Yellow or orange starchy vegetables are particularly healthful. They contain beneficial plant chemicals called carotenoids such as beta-carotene that have "provitamin A activity," meaning that your body produces vitamin A from them. They also protect your cells against damage that could lead to health problems. With their many nutritional "pluses," starchy vegetables are an important component of a healthy diet.

Peas and Legumes

These energy-loaded foods include black beans, black-eyed chickpeas (garbanzos), kidney beans, lentils, lima beans, navy beans, pinto beans, soybeans, split peas, white beans, and even peanuts.

There are 13,000 kinds of legumes, but only about twenty are used as food by humans. Among those twenty, there is a wide variety of colors, sizes, shapes, and flavors. Calorie-wise though, most dry beans are similar. A half-cup serving provides between 110 and 143 calories of energy-giving carbohydrates.

Some legumes, such as soybeans and peanuts, are cultivated for both their protein and their oil content. Others, like lentils, lima beans, chickpeas, and peas, are grown for their value as protein sources. In either case, these foods are among the most nourishing of all vegetables—and are extremely high in fiber. Beans are also packed with vitamins, especially thiamin, riboflavin, niacin, and folate; and minerals, including calcium, iron, copper, zinc, phosphorus, potassium, and magnesium.

Fruits and Vegetables

Fruits and certain vegetables other than legumes are not as high in starch as other carbohydrate foods, but they are loaded with other vital nutrients that protect your body against disease. These include vitamins, minerals, antioxidants, and a variety of natural plant chemicals called "phytochemicals" that have wide-ranging health benefits. All these nutrients are known to guard against cancer, heart disease, and other life-threatening illnesses. Eating a fruit- and vegetable-based diet ensures that you get the greatest variety of disease-fighting nutrients.

An easy way to do that is to make it a habit to eat one or more servings a day from each of these categories of fruits and vegetables: citrus fruits; noncitrus fruits, including berries; green and dark-green leafy vegetables, including spinach and romaine lettuce; yellow/orange or red vegetables such as sweet peppers, carrots, and squash; and "cruciferous" vegetables such as broccoli, Brussels sprouts, or cabbage. Cruciferous vegetables are so named because their reproductive structures contain components that are arranged like a cross, hence the name "cruciferous," which comes from a word meaning "to place on a cross," or "crucify."

Dairy Products

You probably don't think of dairy products as carbohydrates, but these foods do hold an appreciable amount of the nutrient. Most dairy products contain lactose, the only major carbohydrate other than galactose that comes from an animal source. Lactose is important because it stimulates the absorption of calcium from your intestine.

Some people, however, are sensitive to lactose—a condition called "lactose intolerance"—because they lack sufficient lactase, the enzyme required to digest lactose. If you are lactose intolerant, you can't drink milk without getting an upset stomach. If you have problems digesting milk, try using lactase enzyme replacers, such as Lactaid or Dairyease, which can help you add more milk to your diet, or include yogurt in your diet. Yogurt is more easily digested if you're lactose intolerant. Check the label to make sure it contains live active yogurt cultures such as *L. acidophilus*, which can destroy dangerous food-borne bacteria and, thus, offer protection against disease.

Sugars, Natural and Added

Sugar in our diets comes in two forms: naturally occurring sugar and added sugar, or what scientists called "extrinsic sugar." Naturally occurring sugar is found in fruits and vegetables, dairy products, and such sources as sugar cane, sugar beets, and honey. Virtually all the calories supplied by fruit come in the form of sugar.

The most common form of added sugar is what you know as table sugar, which is incorporated into food, baked goods, desserts, candies and sweets, and processed foods. Another type of added sugar is "high-fructose corn syrup," a refined version of fructose made from cornstarch and found in soft drinks, beverages, and many processed foods. You'll learn more about high-fructose corn syrup and other added sugars in the next chapter.

The amount of added sugar in the American diet is on the rise. In 1970, we ate about 120 pounds of sugar per year; today, we eat 150 pounds a year, or almost a half-pound a day. Most of this sugar comes from the soft drinks we consume, as well as from table sugar, jams, and syrups.

Alcohol

Considered a minor carbohydrate, alcohol is formed during a process called fermentation in which sugars and starches are broken down by yeast. Beer, for example, is made by fermenting barley and hops; wine, by fermenting grapes and other fruits; and hard liquor, by fermenting various grains or potatoes. After fermentation, the liquid is distilled to concentrate the alcohol.

Your body uses alcohol as an energy source, but it is not converted to glucose as other carbohydrates are. Instead, it is converted into fatty acids (components of fat). When consumed in excess of what is burned off, alcohol will be turned into fat and stored.

Alcohol provides nothing more than calories to your diet; it contains no protein and very few vitamins or minerals. Although there are some beneficial plant chemicals in beer and wine, you can obtain the very same substances from fruits and vegetables. From a nutritional perspective, alcohol cannot be recommended as a component of a balanced diet and should be used in moderation, if at all.

An Energizing Nutrient

Clearly, carbohydrates are a vital nutrient for good health, supplying the get-up-and-go you need to get up and go, plus providing a bounty of vitamins, minerals, fiber, and other protective nutrients. They are the mainstay of a good diet, provided you learn how to select the healthiest carbs possible. That's where we are headed in the next chapter.

Carbohydrate Scorecards

Carbs are the "enemy" nutrient of the moment, but what is the real story anyway? What carbs should you eat, and what carbs should you avoid? Let's start answering these questions by decoding carbohydrates so you can see which ones work for you nutritionally (the good carbs) and which ones work against you (the bad carbs).

There are essentially three ways to grade carbohydrates to figure out which ones yield the most bang for your nutritional buck. The carbohydrate "scorecards" are as follows:

- Simple versus complex

- Fast versus slow

- Low fiber versus high fiber

Each of these appraisal systems has its pluses and minuses, so it's best to become familiar with all three methods. When you begin to look at carbs from all three dimensions, you will have a sharper understanding of which ones best suit your individual health requirements. Then, you can begin to customize your diet to accommodate your body's true carbohydrate needs.

Right now, the carbohydrate picture might look a little fuzzy to you, but trust me, it will come into much sharper focus by the time you finish this chapter.

Scorecard #1: Simple Versus Complex

Carbohydrates are traditionally categorized as either *simple* or *complex*, a classification that is based on the molecular structure of the carbohydrate, with simple carbohydrates, or sugars, constructed of either single or double molecules of sugar, and complex carbohydrates (starches) made of multiple numbers of sugar molecules. As I pointed out in the previous chapter, simple sugars are found in table sugar, jams, candies, syrups, and processed foods; complex carbohydrates are found in whole grains, cereals, beans, fruits, and vegetables.

The value of grading carbohydrates in terms of simple versus complex lies in the ability to assess their overall nutrient value. If you desire the highest possible nutrition in your diet, you would choose mostly complex carbohydrates, for example. These foods are packed with nutrients, including dietary fiber, which has a long list of impressive health benefits. Fruits and vegetables, in particular, supply vitamins and minerals, two classes of nutrients vital to health. Several vitamins and minerals, namely beta-carotene, vitamin C, vitamin E, and the mineral selenium, are known as "antioxidants." At the cellular level, antioxidants sweep up disease-causing substances known as "free radicals." Free radicals are volatile toxic molecules that cause harmful reactions in the body. In some cases, free radicals puncture cell membranes, preventing the intake of nutrients and, thus, starving the cells. In others, they tinker with the body's genetic material. This produces mutations that cause cells to act abnormally and reproduce uncontrollably. Dreaded diseases like cancer are often the result. In addition to cancer, scientists have linked some sixty diseases to free radicals, among them: heart disease, Alzheimer's disease, and arthritis. Antioxidants guard against free radical damage and are, thus, health-protective. An antioxidant-rich diet— one plentiful in complex carbohydrates—affords significant health protection against these diseases.

By eating lots of antioxidant-rich complex carbs, you're automatically filling up on another set of nutrients called "phytochemicals," which means "plant chemicals." Neither vitamins nor minerals, phytochemicals occur naturally in all fruits, vegetables, and grains. They exert their health-protecting action by various biochemical mechanisms, and the results are nothing short of amazing. Phytochemicals appear to protect against cancer, heart disease, and many other life-threatening illnesses. With their wealth of antioxidants and phytochemicals, you can certainly score complex carbohydrates "good carbs."

Simple carbohydrates, by contrast, are generally lacking in vitamins, minerals, antioxidants, and phytochemicals. Most of the food we commonly describe as "junk food" falls in the simple carbohydrate category. These foods supply a lot of calories but very little in the way of good nutrition. Thus, if your diet is overly high in simple carbohydrates, you're at risk for nutritional deficiencies. By that token, simple carbohydrates can be graded as "bad carbs." For an overview of complex and simple carbohydrates, see the following table.

Who Should Use Scorecard #1?

Answer: everyone. Evaluating carbs as to whether they are simple or complex is the easiest way to make healthy food choices. This scorecard works universally—for weight control, disease prevention and management, and all-around healthy eating.

Scorecard #2: Fast versus Slow

This grading system ranks carbohydrates by how they affect your blood sugar levels. Technically, this system is termed the "glycemic index," and it was invented in the early 1980s by researchers at the University of Toronto. Essentially, the glycemic index is a scale describing how fast a food is converted to glucose in your blood. Foods on the index are rated numerically, with glucose at 100. *The higher the number assigned to a food, the faster it converts to glucose.* Foods with a rating of 70 or higher are generally considered high-glycemic index foods (high-GI foods); a rating of 55 to 69, moderate-glycemic index foods (moderate-GI foods); and below 55, low-glycemic index foods (low-GI foods).

High-GI foods tend to cause a rapid surge of insulin (the hormone that helps usher glucose into cells for fuel), followed by a plunge of blood sugar that can lead to low energy. By contrast, low-GI foods produce a slow, steady release of insulin and yield more sustained energy. Theoretically, you can use the index to select carbohydrates that are less likely to cause roller-coaster swings in your blood sugar.

	EAT MORE OF THESE	EAT LESS OF THESE
	Complex Carbohydrates (Good Carbs)	*Simple Carbohydrates (Bad Carbs)*
Beverages	Freshly squeezed juice Natural juices, no sugar added	Alcoholic beverages Any sugared drink Flavored coffees and teas Fruit punches Juice drinks, sugar added Lemonades Punches Sugared soft drinks Sugared waters
Bread, Grain, and Cereal Products	Whole-grain breads Whole grains and cereals	Cakes, cookies, other baked goods Crackers, chips, and other snack foods (potato chips, taco chips, corn chips, pretzels, cheese curls, etc.) Croissants, muffins, and biscuits Fried starches (tortillas and taco shells) High-calorie convenience foods Pies Pizza dough Presweetened cereals Sweet breads White breads and rolls White pastas White rice

(continued)

	EAT MORE OF THESE	EAT LESS OF THESE
	Complex Carbohydrates (Good Carbs)	*Simple Carbohydrates (Bad Carbs)*
Dairy Products	Fat-free milk Low-fat milk Skim milk Yogurt, sugar free	Chocolate milk Frozen desserts Ice cream Ice milk Milk shakes Puddings, sugar added Soft ice cream Whole milk Yogurt, sugar added
Fruits	Fresh fruits Frozen fruits, no added sugar Fruits canned in their own juice or packed in water	Fruits canned in syrup Fruits to which sugar has been added
Vegetables	Canned vegetables, no added sugar or fat Frozen vegetables, no added sugar or fat Legumes Lightly cooked vegetables Raw vegetables	French fries Instant mashed potatoes Vegetables with added sugar Vegetables with cream sauces

Why is this important? Whether a carbohydrate causes a fast release (high-GI carbs) or a slow release (low-GI carbs) can have dramatic effects on many aspects of your health—all due to insulin and its effects on your body. At normal levels, insulin is a vital hormone, with many important functions in the body. But when repeatedly elevated, insulin is harmful, contributing to diabetes, obesity, heart disease, and possibly cancer.

In diabetes, for example, the glycemic index can help patients select foods that help them better control their blood sugar. Research shows that a low-GI diet lowers blood glucose and insulin levels significantly compared to diets of high-GI foods. Scientists also believe that following mostly a low-GI diet might help prevent diabetes.

Low-GI diets might help you control your weight, too. Remember that in response to high-GI foods, insulin spikes upward. This insulin overload activates fat cell enzymes. These enzymes move fat from the bloodstream into fat cells for storage and trigger your body to create more fat cells. High-GI foods thus create conditions in your body that are conducive to gaining fat.

High-GI foods also stimulate your appetite. These carbs initiate a steep rise in blood sugar, which then drops to subnormal levels about an hour or two after you eat that high-GI meal. This abnormally low level of blood sugar creates a state of hunger. What all this tells us is very simple: High-GI foods promote fat storage; low-GI foods do not. Choosing low-GI carbohydrates makes it possible to lose weight more easily by controlling insulin, fat storage, and your appetite.

Where heart disease is concerned, high levels of insulin can be damaging to your heart. Chronically elevated insulin triggers high blood sugar, elevates triglycerides (a type of blood fat associated with heart disease), and reduces levels of a beneficial type of cholesterol known as HDL.

Then there is the cancer issue. An emerging body of medical research indicates that a high-GI diet might be linked to colon cancer, the third leading cause of disease-related death in the United States. Again, insulin is the villain; it appears to accelerate the development of colon cancer by increasing certain growth factors that lead to uncontrolled cell division.

Now for some caveats: Remember my mentioning that each carbohydrate grading system has its inherent flaws? The glycemic index, despite its usefulness, has a few chinks in its evaluative armor. Here's why: Using the glycemic index to select foods can sometimes be misleading—for at least three reasons. First of all, meals do not consist of single foods. Normally, you eat mixed meals containing protein, carbohydrates, fiber, and some fat.

Mixed meals, thanks to their combination of protein, fiber, and fat, slow your digestion, providing a steady insulin release and, therefore, lowering the glycemic index of the entire meal. The point is, a food's rating on the index does not make much difference as long as it is eaten in slow-release food combinations.

The second problem with this method of rating foods is that some people choose foods low on the index. Many of these foods are simply not good for you, especially if you're trying to control your weight. Take ice cream, for example. It is a low-glycemic index food because it contains protein and fat, along with high-glycemic sucrose (sugar). Thus, it is broken down very slowly. Although it has a low glycemic index, it is chock-full of fat and sugar. Some foods with a low glycemic index can lead to unwanted body fat and other health problems.

On the other hand, potatoes, cantaloupe, tropical fruits, and pumpkin have a relatively high glycemic index, yet are highly nutritious—good carbs, really. If you shunned such foods, you'd be missing out on foods loaded with vitamins, minerals, and fiber. Letting the glycemic index totally dictate the carbohydrates you choose is not always wise.

The third problem with using the glycemic index is that the ratings of foods can change, depending on their individual characteristics as well as how they are prepared. Take a banana, for example. Its rating practically doubles with its ripeness! In addition, cooking starchy vegetables breaks down their cell walls—a process that increases their glycemic score.

Who Should Use This Scorecard?

The glycemic index remains controversial, and not everyone needs to use it. Judging carbs on whether they are complex or simple works best for most people. Still, there is value in using the glycemic index—in certain cases. Consider using this scorecard if you need to:

- Control your blood sugar, particularly if you have diabetes or want to prevent it, especially if it runs in your family.

- Watch your weight.

- Tame an out-of-control appetite (in research, low-glycemic foods have been deemed more filling than high-glycemic foods).

- Control levels of certain blood fats (see chapter 8 for more information).

The glycemic index is also useful in sports nutrition. Low-glycemic foods eaten prior to long bouts of exercise have been found to extend endurance time. High-glycemic index foods eaten after exercise are more effective at rapidly restoring carbohydrates lost from muscles as a result of training or working out.

LOW-GLYCEMIC INDEX CARBOHYDRATES
(LESS THAN 55)

Beverages	*Whole milk*	27
	Soy milk	31
	Fat-free milk	32
	Skim milk	32
	Chocolate milk	34
	Apple juice	41
	Carrot juice	45
	Pineapple juice	46
	Grapefruit juice	48
	Cranberry juice cocktail	52
Bread, Grain, and Cereal Products	Pearled barley	25
	Fettuccine	32
	Whole-wheat spaghetti	37
	Spaghetti	41
	Pumpernickel bread	41
	All-bran cereal	42
	Macaroni	45
	Bulgur wheat	46
	Banana bread	47
	Long-grain rice	47
	Parboiled rice	47
	Old-fashioned oatmeal	49
	Oat bran	50
	Stone-ground, whole-wheat bread	53
	Special K cereal	54
Dairy Products	Low-fat, sugar-free yogurt	14
	Low-fat with sugar yogurt	33
	Low-fat ice cream	50
Fruits	Berries	15
	Cherries	22
	Plums	24
	Grapefruit	25
	Peaches	28
	Canned peaches	30
	Banana, underripe	30

LOW-GLYCEMIC INDEX CARBOHYDRATES
(LESS THAN 55)

Fruits	Dried apricots	31
	Apples	36
	Pears	36
	Grapes	43
	Oranges	43
	Kiwifruit	52
	Bananas, ripe	53
Legumes	Peanuts	14
	Soybeans	18
	Kidney beans	27
	Lentils	29
	Black beans	30
	Lima beans	32
	Split peas	32
	Chickpeas	33
	Lentil soup	44
	Pinto beans	45
	Baked beans	48
Vegetables	Tomato soup	38
	Peas	48
	Yams	51
	Sweet potatoes	54
Miscellaneous	Fructose	23
	Candy bars	41
	Potato chips	54
	Oatmeal cookies	54

*Eat these foods sparingly. Although low on the glycemic index, they are high in empty calories and provide few nutrients; some of these foods are high in fat.

MODERATE-GLYCEMIC INDEX CARBOHYDRATES (55 TO 69)

Beverages	Orange juice from concentrate	57
	Carbonated soft drinks	68
Bread, Grain, and Cereal Products	Brown rice	55
	Linguine	55
	Popcorn	55
	Sweet corn	55
	White rice	56
	Pita bread	57
	Mini shredded wheat	58
	Blueberry muffins	59
	Bran muffins	60
	Hamburger buns	61
	Granola bars, chewy	61
	Couscous	63
	Instant oatmeal	66
	Grape nuts	67
	Croissants	67
	American rye bread	68
	Taco shells	68
	Whole-wheat bread	69
	Cornmeal	69
Dairy Products	*Ice cream*	61
Fruits	Fruit cocktail, canned	55
	Mangoes	55
	Peaches, canned, heavy syrup	58
	Papayas	58
	Apricots, canned, light syrup	64
	Raisins	64
	Pineapples	66
Legumes	Black bean soup	64
	Green pea soup	66

(continued)

MODERATE-GLYCEMIC INDEX CARBOHYDRATES
(55 TO 69)

Vegetables	Beets	64
Miscellaneous	*Table sugar	65

*Eat these foods sparingly. Although relatively low on the glycemic index, they are high in empty calories and provide few nutrients; some of these foods are high in fat.

In the preceding tables, you'll find the glycemic index of some common carbohydrate foods.

Scorecard #3: Low Fiber versus High Fiber

I'm sure you know all about fiber. It's what your grandmother used to call "roughage," and what we know as the stuff in fruits, vegetables, and whole grains that keeps us regular. The "push" response of your intestines depends on adequate fiber in your diet. Low amounts of fiber in the diet are linked to dozens of medical problems, including heart disease, some cancers, diabetes, diverticulosis, and gallstones. See the following table for a list.

Just so you know what carbs to avoid: Low-fiber carbs include such foods as white bread, white rice, cream of wheat, corn flakes, iceberg lettuce, and any simple carbohydrate such as those listed earlier. The more processed and sugary a food is, the less fiber it contains.

Fiber is essential if you want to stay healthy and active as long as possible, so it is important to judge carbohydrates in terms of their fiber content. Generally, food labels will help you make this determination. A food is considered a high-fiber food if it contains at least 2.5 to 3.0 grams of fiber per serving. Without question, high-fiber foods are considered to be good carbs. In the following table, you'll find a list of some of the best.

There are two general types of fiber in foods: soluble and insoluble. Soluble fiber—beta-glucans in oatmeal, pectin in apples, and hemicellulose in beans—dissolves in the presence of water, forming a gummy gel. It is this gel that binds to toxic substances to usher them out of your body. Hemicellulose, in particular, helps lower the amount of fat that is absorbed during digestion.

HIGH-GLYCEMIC INDEX CARBOHYDRATES
(70 OR HIGHER)

Beverages	Gatorade	91
Bread, Grain, and Cereal Products	Melba toast	70
	White bread	70
	Bagels	72
	Kaiser rolls	73
	Bran flakes	74
	Cheerios	74
	Cream of wheat, instant	74
	Graham crackers	74
	Saltines	74
	Doughnuts	75
	Frozen waffles	76
	*Total cereal	76
	English muffins	77
	Rice cakes	82
	Rice Krispies	82
	Cornflakes	84
	Rice Chex	89
	Rice, instant	91
	French bread	95
Fruits	*Watermelons	72
Vegetables	*Carrots	71
	*Rutabaga	72
	*Potatoes	85
	*Parsnips	95
Miscellaneous	Corn chips	72
	Honey	73
	Pretzels	83

Although high on the glycemic index, these are very healthy foods and should not be avoided or limited in your diet.

CONDITIONS AND DISEASES ASSOCIATED
WITH POOR DIETARY FIBER INTAKE

Appendicitis	Gallstones
Cardiovascular disease	Gum disease
Breast cancer	Hemorrhoids
Cholecystitis (inflammation of the gallbladder)	Hiatal hernias
	High blood pressure
Colorectal cancer	High cholesterol
Constipation	Ischemic heart disease
Coronary blood clotting	Kidney stones
Dental cavities	Obesity
Duodenal ulcers	Pyelonephritis (inflammation of the kidney)
Diabetes	
Diverticulosis	Stomach cancer
	Varicose veins

Soluble fiber is also food for the billions of helpful bacteria that reside in your intestines. Good sources of soluble fibers include barley, rice, corn, oats, legumes, apples and pears (especially the fleshy portions), citrus fruits, bananas, carrots, prunes, cranberries, seeds, and seaweed.

Insoluble fiber, which includes lignins, cellulose, and some hemicelluloses, swells in water but does not dissolve in it. Lignins usher bile acids and cholesterol out of the intestines. Cellulose, the roughage we tend to associate with fiber, acts like a stool softener and bulk former, improving elimination and flushing carcinogens from your body. Hemicelluloses do their part by absorbing water in the digestive tract and moving food faster through your system. These actions help relieve constipation, rid the body of cancer-causing substances, and assist in weight control. Good food sources of insoluble fibers include root and leafy vegetables, wheat bran, whole grains (such as wheat, barley, rice, corn, and oats), legumes, unpeeled apples and

HIGH-FIBER FOODS

	Food	Serving Size	Fiber Content (gm)
Beans and Legumes	Beans, kidney, red, canned	½ cup	8
	Peas, split, boiled	½ cup	8
	Lentils, boiled	½ cup	8
	Beans, black, boiled	½ cup	7.5
	Beans, pinto, boiled	½ cup	7.5
	Refried beans, canned	½ cup	6.5
	Lima beans, boiled	½ cup	6.5
	Beans, kidney, red, boiled	½ cup	6.5
	Beans, baked, canned, plain or vegetarian	½ cup	6.5
	Beans, white, canned	½ cup	6.5
Vegetables	Artichokes, boiled	1 cup	9
	Peas, boiled	1 cup	9
	Vegetables, mixed, boiled	1 cup	8
	Lettuce, iceberg	1 head	7.5
	Pumpkin, canned	1 cup	7
	Peas, canned	1 cup	7
	Brussels sprouts, frozen, boiled	1 cup	6
	Parsnips, boiled	1 cup	6
	Sauerkraut, canned	1 cup	6
Breads, Grains, and Cereals	Fiber One (General Mills)	½ cup	14
	Granola, homemade	½ cup	13
	All-Bran with Extra Fiber (Kellogg's)	½ cup	13
	All-Bran (Kellogg's)	½ cup	9.7
	Bulgur wheat, cooked	½ cup	8
	Raisin Bran (Post)	1 cup	8
	100% Bran (Post)	⅓ cup	8
	Bran Chex (Kellogg's)	1 cup	8
	Shredded wheat and bran	1¼ cup	8
	Oat bran, cooked	½ cup	6

(continued)

HIGH-FIBER FOODS

	Food	Serving Size	Fiber Content (gm)
Fruits	Avocadoes, Florida	1 avocado	18
	Raspberries, frozen, unsweetened	1 cup	17
	Prunes, stewed	1 cup	16
	Dates, chopped	1 cup	13
	Pears, dried	10 each	13
	Avocadoes, California	1 avocado	8.7
	Raspberries, raw	1 cup	8
	Blueberries, raw	1 cup	7.6
	Papayas, whole	1 fruit	5.5
	Figs, dried	2 figs	4.6
	Pears	1 medium	3

Source: Nutrient Data Laboratory. USDA Nutrient Database for Standard Reference, Release 14. Beltsville, Maryland.

pears, and strawberries. High-fiber cereals are usually fortified with insoluble fiber.

Fiber Requirements

How much fiber should you eat to get its protective benefits? The current recommendation is 25 to 35 grams of fiber a day from both soluble and insoluble fiber. Most Americans get only about 11 grams a day, however. If you're fiber-needy, try to increase your intake gradually. Doing so can help prevent cramping, bloating, and other unpleasant symptoms often associated with increased fiber. Always drink plenty of water, too—about 8 to 10 glasses of pure water daily—because a high-fiber diet is virtually worthless unless there's enough water in your system to help move the food and fiber through. There are many ways you can sneak fiber into your diet to increase your daily intake, and you will learn about them in this book.

Who Should Use This Scorecard?

Because everyone needs to eat more fiber, everyone can benefit from using this scorecard. Knowing the fiber content of foods and increasing your intake is very important if you need to:

- Manage digestive diseases such as diverticulosis, or prevent constipation.

- Control your weight, as fiber is a top-drawer nutrient for reducing body fat.

- Prevent or treat heart disease or diabetes.

- Protect yourself against cancer and other diseases.

The Scoop on Sugar

Among simple carbohydrates, sugar, or sucrose, is one of the least desirable carbs to include in your diet. Sugar is unfortunately a major source of calories in the diet, but it is a highly refined food that offers no nutrients to go along with the calories. For this reason, dietitians describe sugar as having "empty calories." Therefore, it's best to restrict it in your diet. Too much sugar has been associated with tooth decay, obesity, cardiovascular disease, cancer, and blood-sugar-metabolism disorders such as diabetes and hypoglycemia (low blood sugar). Unfortunately, sugar and other natural sweeteners are added to a dizzying array of foods, but if you stick to natural, unprocessed carbohydrates, you'll automatically slash your intake of added sugar and never go overboard.

High-Fructose Corn Syrup: The Sugar That Acts Like a Fat

One particularly troublesome form of added sugar is high-fructose corn syrup. From soft drinks to cereals to energy bars, foods that you probably eat every day are sweetened with this additive, and it might just be the baddest of the bad carbs.

Made from cornstarch, high-fructose corn syrup is a liquid that is predominantly fructose but has some glucose in it. (Even the powdered fructose sold in stores is made from cornstarch and not from refining the sugar in fruit, as you might assume.) Food manufacturers love high-fructose corn syrup because it tastes much sweeter than sugar; this means they can use less of it and save on their manufacturing costs. Consumption of high-fructose corn syrup has risen more than 21 percent since 1970, when it was introduced into the food supply.

Back then, no one knew that it harbored a secret: High-fructose corn syrup is metabolized differently in the body than regular, good ol' sugar. To help you understand these differences, here's a snapshot of what happens:

When you eat sugar, it is rapidly digested and released into your bloodstream, elevating your blood sugar. This triggers the production of insulin by your pancreas to clear the sugar from your blood and into cells to be used for fuel. Besides insulin, sugar also increases the production of leptin, a hormone that acts like an appetite suppressant. Sugar suppresses the secretion of another hormone, ghrelin. Produced mostly by cells in your stomach, ghrelin triggers your desire to eat. So to a certain extent, eating carbohydrates that contain some glucose helps dampen your hunger.

The metabolism of fructose takes an entirely different course. Surprisingly, fructose appears to act more like a fat in the body: It does not trigger insulin production. It does not increase leptin production. Nor does it suppress the production of ghrelin.

Further, if not used first as energy, fructose bypasses a dietary control point that other sugars go through and instead heads directly to your liver, where it is metabolized into triglycerides, a fat. What all this suggests is that eating a lot of fructose, like eating too much fat, might make you gain weight.

Not surprisingly, many scientists believe that fructose, particularly high-fructose corn syrup because it is so prevalent in foods, is a contributing factor to the obesity epidemic in this country. Its increasing consumption has paralleled the rise in obesity among Americans. According to a U.S. Department of Agriculture analysis, from 1985 to 2000, Americans added roughly 330 calories to their daily intake, and 25 percent (about 83 calories) came from sweeteners, including high-fructose corn syrup. That amount of added sugar in your diet will produce a weight gain of nearly nine pounds a year.

Fructose has another downside: Some people suffer from a condition called "fructose intolerance"—a sensitivity to the fructose in fruit juices,

sports drinks, or products containing high-fructose corn syrup, and sometimes to the natural fructose in fruit. Symptoms include stomach upset, diarrhea, and bloating. In rare cases in which someone lacks the enzyme needed to
digest fructose, the symptoms will be more severe: vomiting, hypoglycemia
(low blood sugar), jaundice, or enlarged liver. You can be tested for fructose
intolerance with a breath test or a blood test. If diagnosed with this condition,
you'll have to avoid foods containing fructose, especially high-fructose corn
syrup.

Another fructose alert: Research with fructose has discovered that high
levels of fructose in the diet can keep the body from properly using magnesium and copper, two essential minerals.

Hidden Sugars

Go on alert that you might be eating more sugar than you think you are
due to hidden sugars and sweeteners. A food could still contain simple sugars but under the guise of names like honey, molasses, corn syrup, and, yes,
high-fructose corn syrup. The World Health Organization (WHO) now recommends that you limit your intake of added sugars to 10 percent of your
total daily calories. So if you eat 2,000 calories daily, this works out to 200
calories, the amount found in a slice of apple pie, a half-cup of ice cream, or
a can and a half of cola.

In the following table, you will find many of the different forms of sugars found in food. If these ingredients appear first or second on a label's list
of ingredients, the product is too high in added sugars. Be a wise consumer.
Make sure you know what you're buying—and what you're eating.

When Good Carbs Go Bad

Recognize that any good carb can be degraded into a bad carb through processing; through the addition of sugar, fat, or additives; or through overcook-
ing. A good example is the potato. What starts out as a highly nutritious,
filling, nutrient-rich carb when baked, morphs quickly into a miserable,
greasy, high-calorie, bad carb when made into french fries or potato chips.
The point is, eating mostly good carbs—foods that are natural and
wholesome—will help keep you in peak health.

HIDDEN SUGARS IN YOUR FOOD

Sugar (sucrose)	Refined crystallized sap of the sugar cane or sugar beet; a combination of glucose and fructose.
Dextrose (glucose)	Another name for glucose. A simple sugar that is less sweet than fructose or sucrose.
Lactose	A simple sugar from milk; less sweet than sucrose, fructose, or maltose.
Maltose	A simple sugar made from starch; less sweet than sucrose or fructose.
Maltodextrin	A manufactured sugar from maltose and dextrose in corn.
Brown sugar	A refined sugar coated with molasses. The minerals calcium, iron, and potassium are present in this sugar.
Raw Sugar	A less-refined sugar that still has some natural molasses coating.
Fructose	A simple sugar refined from fruit.
Molasses	The syrup separated from sugar crystals during the refining process. Blackstrap molasses, a popular health food, is a good source of calcium, iron, and potassium.
Honey	A concentrated solution of fructose and glucose (80 percent) and some sucrose. Produced by bees from the nectar of flowers.
Maple syrup	A concentrated sap from sugar maple trees, predominantly fructose.

(continued)

HIDDEN SUGARS IN YOUR FOOD

Corn syrup	A manufactured syrup of cornstarch, containing varying proportions of glucose, maltose, and dextrose.
High-fructose corn syrup	A highly concentrated syrup of mostly fructose and some glucose; very prevalent in soft drinks and other processed foods.
White grape juice	A highly purified fructose solution; virtually no other nutrients are present.
Mannitol, Sorbitol, Xylitol	Sugar alcohols derived from fruit or produced from dextrose. All three contain the same number of calories as sugar (4 calories per gram). Sugar alcohols are more slowly absorbed than sugar is. For this reason, the sugar alcohols are thought to be better than sugar for diabetics. Sugar alcohols do not promote tooth decay, unlike sugar.

Summing Up

Carbohydrates can be evaluated in at least three different ways, all useful in their various applications. As you study the three scorecards, you can see that good carbs share certain characteristics, regardless of the scorecard applied. Good carbs are:

- Plentiful in vitamins, minerals, antioxidants, and phytochemicals.
- High in disease-preventing fiber.
- Free of added sugar and harmful additives.
- Pure, unrefined, and minimally processed.
- Generally helpful in controlling blood sugar and insulin levels.
- Protective against an array of diseases.

As you make your daily food choices, keep these characteristics in mind. Your body prefers good carbs and uses them so much more efficiently than it uses processed, chemical-laden bad carbs, such as refined cereals, commercially baked goods, or fat, sugar, and additive-loaded packaged foods. Bad carbs are nutritionally bankrupt and associated with various health problems. Good carbs are bursting with nutrients, each put to use in building and healing the body.

· THREE ·

Healthy Bites:
The Twenty-five Super Carbs

Without question, carbohydrates are a vital food for supplying energy and supporting good health. And as you have seen, some carbs are more desirable than others. But even among those good carbs, some deserve to go to the head of the line, because they are brimming with natural chemicals and other nutrients that are exceptionally strong guardians of health.

Here are twenty-five super carbs to stock in your kitchen in order to boost your immune system, protect yourself against disease, and perform at your very best. Each is a winner on all three carb scorecards discussed in the previous chapter. Make it a habit to serve up a variety of these foods, along with other good carbs, every day of the week.

1. Apple

There's a reason why the old saying "an apple a day keeps the doctor away" has had such a long life: Apples have a slew of health benefits all attributable to their high phytochemical and high fiber contents.

One of the major phytochemicals in apples is quercetin. Research indicates that quercetin deters the changes that make prostate cells cancerous and might retard the cancer's spread. It also might reduce the risk of lung cancer. A 25-year follow-up study involving nearly 10,000 Finnish men showed a reduced risk for lung cancer in cases where the consumption of apples was high. In animals, quercetin has also been found to inhibit the growth of melanoma, a deadly form of skin cancer.

Apples are high in fiber, too, with one fruit supplying 5 grams of fiber. Roughly 80 percent of that fiber is soluble and, thus, heart-protective; the rest is insoluble, the form believed to exert an anticancer benefit. You'll also find plenty of vitamin A, B vitamins, and various minerals in apples.

An apple a day keeps the dentist away, too. When you crunch down on an apple, it scours away some of the bacteria and plaque that adheres to your teeth.

2. Artichokes

How fondly I remember my favorite lunch as a child—boiled artichokes, their tender leaves dipped in my mother's homemade mayonnaise. I'd pull the leaves between my teeth (there's an "art" to eating artichokes) and feast on their delicious pulp.

Little did I know then that I was dining not only on a delectable vegetable but also on one whose leaves are chock-full of phytochemicals. Among the phytochemicals in artichoke leaves are flavonoids, plant pigments that act as antioxidants; fructo-ogliosaccharides (FOS), a group of beneficial nondigestible carbohydrates; caffeoylquinic acids, which save cells from free radical damage; and luteolin, an antioxidant that appears to reduce levels of LDL (bad) cholesterol.

Artichoke (*Cynara scolymus*) is a member of the aster family, to which daisies, dandelions, chrysanthemums, and the milk thistle plant also belong. What you see in the supermarket is actually the flower of this plant.

Since ancient times, artichoke leaf has been used medicinally as a "choleretic" (bile increasing) and as a diuretic. In folk medicine, it is a well-known cure for digestive problems. Over the past few decades, mounting scientific evidence has verified the validity of these traditional claims, plus it has shown the artichoke leaf to be a bonafide treatment for other conditions, namely high cholesterol and liver problems.

If you've never prepared an artichoke, it's very easy to do. Simply boil the artichokes in a large saucepan of water (add 2 tablespoons vinegar for flavoring) for about 45 minutes. Dip the leaves in your favorite low-fat or reduced-fat salad dressing, and enjoy. Canned artichoke hearts make great additions to salads.

3. Avocado

Okay, I know you're thinking that an avocado is a fat and, yes, it is extremely high in *healthy* fats. But the avocado, technically a fruit, is a mixture of protein, fat—and carbohydrate—so it's difficult to pigeonhole it into any one nutrient category. There are two varieties of avocado—Florida and California—and both are equally loaded with nutrients.

Avocados are also a valuable source of fiber. The California avocado yields 8 grams of fiber; the Florida variety, 16 grams. Their antioxidant content is second to none. Avocados contain glutathione (more than three times that found in other fruits), and are the highest fruit source of lutein. Glutathione detoxifies cancer-causing agents. Lutein is best known for preventing eye diseases, and it might guard against cancers of the colon, lung, and breast. Avocados also contain more vitamin E than any other fruit.

People used to shun the avocado because of its fat, but now we know better. The fat in this fruit is of two major types: monounsaturated, which is known to keep cholesterol levels in check, and beta-sitosterol, also a good cholesterol regulator.

4. Beets

If you're like most people, beets probably aren't a mealtime staple at your house, although they should be. This root vegetable is a nutritional treasure trove. A cup of cooked sliced beets, for example, yields 3 grams of fiber. Plus, beets are loaded with disease-fighting antioxidants. Also going for beets: One serving of beets provides more than a third of your daily requirement for folate. You probably know this nutrient best as folic acid, but folate is the form found in food; folic acid is the supplement form. Folate, a member of the family of B vitamins, is critical for the synthesis, repair, and functioning of DNA, the genetic material of cells. Folate is a super-critical nutrient that might reduce your risk of heart disease, stroke, and colon cancer.

5. Berries

Blueberries, strawberries, raspberries, and blackberries are among the healthiest fruit carbohydrates you can eat because they're loaded with antioxidants that protect you from heart disease, cancer, and many other diseases. One of the most protective substances in berries is a phytochemical called ellagic acid, which prevents toxic chemicals from damaging cells. Not only that, it helps initiate a process that causes cancer cells to commit suicide.

Blueberries, in particular, have been identified as a fruit that enhances age-related short-term memory, says a Tufts University study conducted with rats. Much like people, rats become more forgetful as they get older, unable to find their way through mazes they once knew how to navigate. But when they were fed extracts of blueberries for two months, they actually improved their ability to navigate through mazes! Not only that, the rats' balance, coordination, and running speed improved. In similar tests, strawberries worked, too, but not as well. Scientists speculate that antioxidants in blueberries reduce inflammation, a process that might harm brain tissue as we get older.

6. Broccoli Sprouts

Although broccoli is certainly a highly nutritious carbohydrate, its forerunner, broccoli sprouts, are even more healthful. Several years ago, scientists at Johns Hopkins University in Baltimore discovered that, cup for cup, broccoli sprouts have 10 to 100 times more anticancer substances in them than regular broccoli. These substances are phytochemicals called isothiocyanates. They work by activating the antioxidant glutathione, which detoxifies carcinogens.

Both broccoli and broccoli sprouts contain a phytochemical called sulforaphane, which also protects cells against cancer. Sulforaphane has another bonus: It appears to eradicate a bacteria in the stomach called *H. pylori*, which causes ulcers.

If you want to start reaping the benefits these sprouts offer, eat about a cup and a half a day. That's the recommended amount for good health. Don't forget to eat your broccoli, too. With its fiber, B vitamins, beta-carotene, and vitamin C (one of the most important antioxidants and cancer fighters), it's worth piling on your plate.

7. Brown Rice

When you want some rice, one of your best choices is brown rice. Compared to white rice, which has been stripped of nutrients, brown rice has its bran layer still intact and is, thus, a top source of B vitamins, including thiamin (vitamin B_1); the minerals calcium, phosphorus, and iron; and fiber. Brown rice is also a good source of vitamin E.

8. Bulgur Wheat

Less familiar among most good carbs, bulgur wheat is a form of whole wheat that can be eaten as a breakfast food or as a delicious side dish at lunch or dinner. You can cook bulgur wheat as you do rice or simply soak it in water or broth. Bulgur is often used as a meat extender or meat substitute in vegetarian cooking. It is best known as the chief ingredient in tabouli salad, a popular Mideastern dish. Nutritionally, bulgur wheat is high in insoluble fiber (5 grams per serving) and is loaded with B vitamins, phosphorus, zinc, copper, iron, and magnesium.

9. Cabbage

The world's oldest cultivated vegetable, cabbage has today become known as an anticancer vegetable. It is high in the cancer-fighting phytochemical glutathione, as well as vitamin A, calcium, and fiber—all nutrients that appear to have cancer-protective effects, particularly in lowering the risk of stomach, breast, and intestinal cancers.

10. Concord Grape Juice

This form of grape juice has more antioxidants than any other juice. In studies, it helps thwart the oxidation of artery-damaging LDL cholesterol—a process caused by free radicals. When LDL cholesterol is oxidized by free radicals, white blood cells in artery linings start attracting excessive amounts of LDL. The oxidized LDL forms fatty streaks on the inner arterial walls, and these streaks become the foundation of atherosclerosis, the abnormal thickening of the arteries that leads to heart disease. There are many different kinds of grape juice on supermarket shelves, and few offer the natural health protection of Concord grape juice. Be sure to look for a brand that specifies that it is 100 percent Concord grape juice.

11. Garlic

In ancient times, garlic was believed to scare off vampires, but now we know that it wards off a great deal more. This centuries-old remedy might have preventive and curative properties in a wide range of ailments, including heart disease, cancer, and immune system disorders. Garlic, it seems, contains more than 200 biologically active compounds that appear to positively alter the course of many illnesses. It is also an antibacterial and an antifungal agent. The best way to secure garlic's healing power is to mash it or mince it prior to adding it to recipes.

12. Greens

There is a long list of green leafy vegetables that are phenomenal in the health benefits they confer. Beet greens, collard greens, dandelion greens, kale, spinach, and turnip greens are just a few of the names on this all-star list of nutritional heroes. What they have in common is their wealth of beta-carotene, vitamin C, folate, iron, and other minerals. These foods also contain phytochemicals that help protect your eyes. Make it a point to eat more green leafy vegetables every day—in salads, soups, or stews.

13. Kiwifruit

Used to be, kiwifruit was relegated to garnish status for fruit trays, but not anymore. Kiwifruit has been promoted to the position of super fruit—and for good reason. This fuzzy little green fruit is packed with vitamin C (more than double what you find in an orange). In fact, one medium kiwifruit provides 120 milligrams of vitamin C; that's twice the daily requirement. For weight control, few fruits beat the kiwifruit. One fruit has just 46 calories.

14. Legumes

Technically an edible seed with a pod, legumes represent a huge family of nutritious vegetables, including beans, peas, and lentils. These vegetables are near-perfect foods, too, containing not only a lot of complex carbohydrates, but plenty of protein as well. As such, they can be used as a meat substitute in vegetarian diets.

Legumes are among the carbs highest in fiber, with 7 grams on average per cup. The soluble fiber contained in legumes helps lower cholesterol. A tip: Don't toss out the liquid in canned beans. Use it for soups, because there is soluble fiber dissolved in this liquid. Legumes are also loaded with iron and the B vitamins folate, thiamin, riboflavin, and niacin.

15. Oatmeal

Oatmeal might not be the most exciting breakfast cereal to wake up to, but it certainly is one of the most healthy. Oatmeal contains a soluble fiber called beta-glucan, which, like a nutritional housecleaner, sweeps out cholesterol-forming particles from your body before they can turn into full-blown artery-clogging cholesterol. Eating oatmeal on a regular basis has been proven to naturally lower dangerous cholesterol and reduce blood pressure. Oatmeal is also one of the few foods that is naturally rich in vitamin E.

Not all oatmeal is alike when it comes to nutritional value, however. Rolled oats, old-fashioned oats, and steel-cut oats are nutritionally superior to the quick or instant varieties.

16. Onions

Yes, onions stink up your breath, but from a health perspective, the stench is worth it. This sometimes unpopular vegetable is beneficial for heart health. It prevents arteries from clogging and contains beneficial substances that stop the formation of dangerous blood clots, which can lead to heart attacks.

Onions contain phytochemicals called saponins, which prevent cancer cells from multiplying, and allylic sulfides, which usher carcinogens from your body, decrease tumor reproduction, and fortify your immune system. Onions are also high in quercetin, an anticancer phytochemical.

17. Oranges

Oranges and other citrus fruits are a storehouse of nutrients, from phytochemicals to vitamins. Best known for their vitamin C content, these fruits also contain phenolics, which are phytochemicals that neutralize carcinogens and stimulate the production of cancer-fighting glutathione. Oranges are a known source of limonene, a phytochemical that stops damaged cells from the uncontrolled growth that can lead to cancer. Another protective nutrient found in oranges is folate, a B vitamin.

Citrus fruits are high in substances called flavonoids, found abundantly in vegetables, grains, tea, and wine. More than 5,000 flavonoids have been discovered in nature, and many are responsible for the bright colors of the fruits and vegetables you eat. Flavonoids are also essential for the proper absorption of vitamin C. In fact, flavonoids are what make natural vitamin C (found in foods) more effective than synthetic supplemental vitamin C by improving and prolonging the function of the vitamin.

18. Pineapple

Probably the world's most favorite tropical fruit, pineapples pack a huge nutritional wallop. They're high in vitamin C, fiber, and the mineral manganese, which protects both your bones and your heart. This delicious fruit is chock-full of a natural enzyme called bromelain, a known anti-inflammatory and a "mucolytic," which means it eliminates the abnormal accumulation of mucus in tissues. Indeed, bromelain has wide-ranging health benefits: It appears to alleviate diarrhea by healing the mucosa of the colon, help prevent heart disease by interfering with the formation of abnormal clots, and possibly stave off cancer due to an antitumor effect. To capitalize on the benefits of pineapple, it's best to eat it fresh, because the processing required to can the fruit destroys most of its bromelain.

19. Red Pepper

When scientists at Cornell University analyzed the antioxidant activity of ten common vegetables in their lab a few years ago, they discovered that red peppers had the highest activity, followed by broccoli, carrots, spinach, cabbage, yellow onions, celery, potatoes, lettuce, and cucumbers. In addition, red peppers were among the top three vegetables found to interfere with the proliferation of human liver cancer cells that had been isolated in a lab dish. Spinach performed the best, followed by cabbage, with red peppers in third place. What this study hints at is that red peppers are a vegetable with much curative and preventive powers.

20. Romaine Lettuce

When you select a lettuce with dark green leaves, such as romaine lettuce, you're getting a real nutritional bargain. Romaine lettuce, the mainstay of Caesar salads, is an excellent source of vitamin C, vitamin A, and folate. Further, it has eight times more beta-carotene and twice as much calcium as iceberg lettuce.

21. Sweet Potato

If you had to eat just one vegetable, it might be a good idea to choose the sweet potato. Ounce for ounce, it is one of the most nutritious vegetables around and loaded with beta-carotene, which converts to vitamin A in your body. Vitamin A is an important antioxidant that strengthens your immune system against bacterial and viral diseases as well as cancer. Sweet potatoes also supply cholesterol-lowering fiber, vitamin C, vitamin E, and the B vitamin thiamin. Although sweet, these spuds are not at all high in calories, providing roughly 120 per medium-size potato.

22. Tomato

Pass the ketchup, the spaghetti sauce, and the tomatoes—please! Tomatoes and tomato products are the primary sources of lycopene, a powerful carotenoid that works forcefully as an antioxidant fighting off disease-causing free radicals. Lycopene appears to be highly protective against cancers of the colon, bladder, and pancreas, but it is particularly noteworthy for its role in preventing prostate cancer. In a diet study sponsored by the National Cancer Institute, researchers identified lycopene as being extremely powerful against prostate cancer. Those individuals who consumed greater than ten servings of tomato-based foods per week had a significantly decreased risk of developing prostate cancer when compared to those who ate fewer than 1½ servings per week. Tomatoes are also a rich source of vitamin C and many other phytochemicals.

23. Wheat Bran

Sprinkling a few tablespoons of raw wheat bran on your cereal in the morning does a world of good for your digestive system. Wheat bran acts as a natural laxative, promoting regularity. For women, there's an added plus of eating wheat bran: It helps move excess estrogen from the body. That's important, because too much estrogen in the body is a risk factor for breast cancer.

24. Winter Squash

Acorn, butternut, hubbard, and banana squash are all well-known members of the winter squash family. Like most orange-, yellow-, and red-colored vegetables, winter squash is high in key carotenoids, including alpha-carotene, beta-carotene, lutein, and zeaxanthin.

Alpha-carotene shows promise in stalling the growth of certain malignant tumors and might be protective against breast cancer. Butternut and hubbard squashes are good sources of alpha-carotene. Beta-carotene reduces the risk of cancers of the colon, rectum, breast, uterus, prostate, and lung. All varieties of winter squash contain significant amounts of beta-carotene.

Lutein, better known for preventing eye diseases, might guard against cancer of the colon, lung, and breast. Its less-well-known companion carotenoid, zeaxanthin, is linked to a lower risk of breast, cervical, and colon cancers. Both carotenoids are being investigated for their role in preventing skin cancer. Acorn squash, which is also very high in fiber, contains lutein and zeaxanthin.

25. Yogurt

Yogurt is produced by fermenting milk with a mixture of bacteria and yeasts. The process yields not only a delicious, custardlike treat, but also a food that is brimming with "probiotics" (a term that means "favoring life"). Probiotics are healthy bacteria that help maintain the health of the digestive tract, preventing the growth of yeast, salmonella, E. coli, and other nasty germs. Among the most well-known probiotic is L. acidophilus, which is plentiful in yogurt. In numerous studies, it has demonstrated antitumor activity, particularly against colon cancer. And, in at least one study, it prevented recurring tumors in bladder cancer patients.

The healthy bacteria in yogurt also have an immune-boosting effect. In a study conducted at the University of California–Davis, researchers found that people who ate 2 cups of yogurt a day had fewer colds.

Yogurt is also rich in bone-building calcium, B vitamins, vitamins A and D, and protein. It is helpful in treating high cholesterol, digestive disorders, and kidney problems.

Footnote: Be sure to avoid yogurts sweetened with added sugar. The best option is to buy plain yogurt and sweeten it yourself with low-sugar or nonsugar jam.

Variety Is the Spice of Life

Along with an active lifestyle, one of the ways to get a superfit body is by eating a diet that includes these super carbs as well as other good carbs. Doing so ensures that you get the greatest variety possible of health-building nutrients. Just remember: Every time you pile these good carbs on your plate, you're serving up a lot more than meets the eye.

Part II

Your Good Carb Prescription for Health

· F O U R ·

Your Brain on Carbs

We don't often think of food as a drug, but food does indeed have druglike effects on the body. It is within the brain that food, particularly carbohydrates, is used to manufacture important brain chemicals called neurotransmitters that, in their ebb and flow, govern your mind, your memory, and even your very behavior in ways that might be quite dramatic. This whole notion that carbohydrates can affect your brain and behavior as powerfully as a drug came to national prominence in the late 1970s, when San Francisco supervisor Dan White claimed that a junk-food diet of Twinkies and Coca Cola put him into a heightened emotional state and deepened his depression when he shot and killed the mayor and another city employee. The defense worked. The jury found White guilty, but of the lesser charge of manslaughter and sentenced him to six years in prison. The celebrated case has since come to be known as the "Twinkie Defense."

To a great degree, the foundations of good, stable mental fitness rest on what you eat. A healthful diet, devoid of highly processed foods, supplies your brain with the nutrients and energy it needs for optimal functioning and psychological health. And certainly, eating too many high-sugar junk foods on a regular basis can be mentally detrimental. Sugar drives your blood sugar up quite rapidly, but this reaction is followed by a fast crash downward as your blood sugar plummets. Low blood sugar, or hypoglycemia, has a depressant effect on the body; it makes you feel blue and out of sorts. However, the association between criminal behavior and the consumption of refined sugar (the Twinkie Defense) has never been verified.

In the information that follows, you'll learn why eating certain carbs (even a little bit of sugar), in the right amounts, holds some of the most powerful secrets to exceptional mental fitness.

Carbohydrates: The Brain Fuel

From the cereal you eat for breakfast to the baked potato you have for dinner, carbs are the leading nutrient fuel for your brain. Your brain uses glucose from the breakdown of carbohydrates almost exclusively in regulating everything mental, from learning to memory to mood. Without ongoing replenishment of glucose, your brain would be deprived of glucose in a mere ten minutes, and your mental power would suffer greatly.

As glucose circulates in your bloodstream, ready to be taken up by cells for energy, your brain gets first dibs on it before any other organ does. And if for some reason glucose is in short supply, your brain will commandeer processes that convert other nutrients into glucose. The health and performance of your brain depend on a readily available supply of glucose, and you can boost your brainpower enormously by adequately fueling yourself with carbs. Carbs play a central role in alertness, concentration, mood, and memory, as well as in protection against a number of brain-damaging diseases. (Some of the important effects of carbohydrates on your brain are listed in the box on the next page.)

Mind Power and Mental Concentration

Maybe you might have to be on your mental toes for an important meeting. Maybe you are scheduled to take an exam. Maybe you are set to learn a new task on your job. Or maybe you're driving long hours toward a destination. Whatever the situation, you need needle-sharp focus and quick response time. You'll be gratified to know that carbohydrates are your knight in shining armor in this regard.

Carbohydrates are powerful nutrients for boosting alertness and concen-

EFFECTS OF CARBOHYDRATES
ON YOUR BRAIN

* Fuels the brain for mental activity

* Increases the production of important brain chemicals such as serotonin

* Produces feelings of calmness

* Reduces anxiety

* Induces sleep

* Heightens recall

* Suppresses the appetite

tration, as long as you select the right ones and in the right amounts. When you are learning a new task, your brain begins to rapidly take up glucose from circulation. Scientists believe that, during a learning activity, glucose may activate the release of a neurotransmitter called acetylcholine to help retain memory. Manufactured from the B vitamin choline, this neurotransmitter is involved mostly in learning and memory. So if you must stay alert through the day, make sure you're fueling your brain with at least one or two servings of good carbs at each meal.

If some of these carbs are rich in vitamin C, you'll boost your alertness even more. Vitamin C increases levels of the neurotransmitter norepinedrine in your brain. Norepinedrine helps you stay alert and motivated. Good choices include citrus fruits, strawberries, kiwifruit, and sweet peppers.

If your powers of concentration need rejuvenating, here's a menu filled with alertness-boosting foods, including carbs, along with information on why they work. (Foods marked by an asterisk indicate that the food is a carbohydrate.)

The Alertness Diet

Breakfast

2 scrambled eggs

*1 slice whole-wheat bread with 1 pat butter or margarine

*1 cup Concord grape juice

1 cup coffee

Mid-Morning Snack

*1 medium apple

Lunch

4 ounces tuna salad (2 to 3 cups mixed *salad vegetables, including 1 *tomato, cut into wedges)

2 tablespoons reduced-fat French dressing

1 cup coffee

Mid-Afternoon Snack

*1 medium orange

2 tablespoons peanuts

Dinner

4 to 5 ounces grilled lean beef

*1 cup cooked mixed vegetables

*½ acorn squash

*1 cup skim milk

Why It Works

- The apple, peanuts, and tomato are loaded with boron. Research indicates that people given additional boron score higher on tests that measure attention and memory.

- The whole-wheat toast is an excellent source of glucose.

- The orange is high in vitamin C, which helps synthesize norepinedrine.

- The lunch in this menu is low in carbohydrates. Here's the reason why:

Too many carbs at lunch impairs attention span and reaction time. To increase alertness, stick to high-protein lunches. Have coffee with lunch if you want to get rid of post-lunch mental slumps. Coffee is a source of caffeine, which boosts alertness.

- The large salad at lunch contains potassium. If you don't get enough of this mineral in your diet, you could have trouble concentrating.

- The grape juice, acorn squash, and other vegetables are chock-full of brain-protecting antioxidants. Concord grape juice, in particular, is one of the most antioxidant-rich foods you can eat.

- The grilled lean beef supplies iron, a mineral important to brain function because it boosts concentration and enhances learning.

Carbohydrate Mood Boosters

Many experts believe that you can defeat a bad mood with what you eat, as long as you select the right foods—those that have a positive effect on brain chemistry. Carbohydrates just happen to be a key "nutrient tranquilizer." If you'd love to soothe jangled nerves, chase away the blues, or lift your spirits, look to carbs for a mood-boosting remedy. A diet rich in carbohydrates can help you feel relaxed and less stressed out, because carbohydrates are indirectly involved with elevating a neurotransmitter called serotonin. Serotonin is known as the "happiness neurotransmitter" because it is associated with tranquility, calm, and emotional well-being.

Eating carbohydrates kicks off a cascade of events that produce serotonin. This process starts when you have a meal that contains protein. Protein foods such as milk and poultry are rich sources of the amino acid tryptophan, a building block of serotonin.

Protein foods contain other amino acids besides tryptophan and in larger amounts. To reach the brain, these amino acids must cross the "blood-brain barrier," a protective network of tightly knit cells lining the blood vessels of the brain. These cells are trusty, vigilant receptionists that screen substances for entry into the brain and bar the door to unwelcome toxins. Those substances that do get in are ferried across the blood-brain barrier by special carrier molecules. Like a passenger vying for a seat on a crowded bus, tryptophan has to compete with five larger amino acids for a ride over. Consequently, not much tryptophan enters, so very little serotonin is synthesized in the brain in response to a high-protein meal.

However, if carbohydrate foods are eaten with protein foods, the carbohydrate helps deliver more tryptophan to the brain. Carbohydrate triggers the release of insulin, which drives amino acids right into brain cells. Thus, eating high-carbohydrate meals—rather than high-protein meals—ships tryptophan into the brain, where it can be eventually converted into serotonin to boost your mood.

To sum up, the process works like this: A high-carbohydrate meal triggers the release of insulin, and more tryptophan can then enter the brain. Tryptophan is used to make serotonin, and the result of more serotonin is a feeling of calmness.

Incidentally, if you crave carbohydrates, it could be because your brain is crying out for more serotonin. (For an explanation of why you crave carbs, see the box on the next page.)

One of the best mood-boosting carbs you can eat is whole-grain bread. Its attraction lies in its mixture of carbohydrates and amino acids, a combo that allows the most efficient delivery of tryptophan to the brain. In fact, whole-grain bread is a near-perfect blend of proteins and carbohydrates that practically ensures that your brain will get enough tryptophan to manufacture serotonin.

Another great mood-boosting carb is the banana. Bananas are well endowed with magnesium, a mineral depleted by stress. When you're chronically stressed out, your body starts churning out more stress hormones, high levels of which cause magnesium to be flushed from cells. This can lead to all kinds of problems, including vulnerability to viruses and mood-sapping fatigue. Bananas are also loaded with tryptophan, the amino acid that improves serotonin metabolism.

Although generally dubbed a "bad carb," particularly in excess amounts, pure sugar is a natural upper in one important regard. Its sweet taste—the sensation you feel when you eat it—triggers the release of feel-good chemicals in your body called endorphins. Researchers have actually observed that the moods of depressed patients brightened considerably when they were fed high-sugar meals. Think twice, however, before popping sugar cubes for a high. A sugar overload will lead to weight gain, and putting on weight is a depressing predicament for plenty of people. Plus, it can lead to mood-deflating hypoglycemia (low blood sugar).

DO YOU CRAVE CARBOHYDRATES?

You know the feeling, that I just gotta have a doughnut . . . a chocolate candy bar . . . some Ben & Jerry's ice cream. Technically, this overpowering desire is called "carbohydrate craving," and it has a biochemical basis. Usually, we start craving carbs because our brain is putting in an order for more serotonin precursors, or building blocks. The chief building block for serotonin is the amino acid tryptophan, available from protein but shuttled into the brain by insulin, which is triggered by eating carbohydrates. So the cravings are essentially a shout for more tryptophan.

Some people are more likely than others to get carbohydrate cravings. They include:

• Those who are overweight or obese. People who are obese often suffer from "insulin resistance," the inability of the body's cells to properly use insulin to drive various substances into cells, including tryptophan. Consequently, not much tryptophan gets shipped into the brain to make serotonin. Trimming down to a smaller size through exercise and diet will resolve insulin resistance and, with it, carbohydrate cravings.

• Women with premenstrual syndrome (PMS). A woman's craving for sweets can intensify during the premenstrual phase of her monthly cycle—a reaction that might also be attributed to fluctuating levels of serotonin in the brain. A highly craved food during this time is chocolate—and yes, it works to ease the depression and anxiety associated with PMS. Chocolate helps synthesize serotonin, possibly because it is high in sugar or some other unknown chemical. Rather than munch on chocolate or other sweets, it's preferable to reach for good carbs such as whole-wheat bread, pasta, or cereals to soothe the premenstrual blues. One study found that eating a carbohydrate-rich, low-protein dinner improved depression, tension, anger, sadness, fatigue, and alertness among a group of women suffering from PMS.

• People with seasonal affective disorder (SAD), a form of depression that strikes 10 to 25 million Americans, usually in the fall and winter. This extreme form of the wintertime blues is thought to stem from abnormalities in the brain chemicals. Low levels of light in the winter disturb the balance of brain chemicals being released, particularly serotonin. Treatments for SAD generally include bright light therapy, exercise, and antidepressant therapy to restore normal levels of serotonin in the brain. Appropriate treatment helps alleviate the cravings.

Many of the good carbs listed in this book are packed with B vitamins, which play a critical role in brain function, from manufacturing neurotransmitters to releasing energy in your brain cells. Two B vitamins of note are folate and thiamin. Found in orange juice, green leafy vegetables, and fortified bread and breakfast cereals, folate has been shown in research to be effective for easing depression. To get the mood-boosting benefits of folate, try to eat a cup of spinach or other green leafy vegetable, several times a week.

Nicknamed the "morale vitamin" because of its beneficial effect on mental attitude, thiamin is plentiful in good carbs such as wheat germ, bran, brown rice, and whole grains. Not only does this important B vitamin help alleviate depression, it can also improve your capacity to learn and grasp new information.

Besides the mood-lifting carbs already discussed, others include corn, dry cereals, green leafy vegetables, muffins, oatmeal, pasta, potatoes, and rice. Here's a sample menu that might help overturn a down-in-the-dumps mood. It is high in good-mood carbs. (Foods marked by an asterisk indicate that the food is a carbohydrate.)

The Blues-Banishing Diet

Breakfast

*1 slice whole-wheat toast

½ cup cooked *oatmeal, sprinkled with 1 tablespoon *bran and 1 tablespoon *wheat germ

*½ cup skim milk

*½ grapefruit

Mid-Morning Snack

*1 cup fruit-flavored yogurt

Lunch

*Vegetable pasta: 1¼ cup pasta topped with 1 cup assorted vegetables and ¾ cup marinara sauce

1 small serving *spinach salad with 1 tablespoon ranch dressing

Mid-Afternoon Snack

*1 granola bar

*1 banana

Dinner

4 to 5 ounces grilled salmon

*½ cup brown rice, with *½ cup red beans

*½ avocado with 1 tablespoon Italian dressing

Bedtime Snack

*½ cup frozen yogurt or sherbet

Why It Works

- The breakfast featured here is high in carbohydrate, which increases brain levels of serotonin, the neurotransmitter responsible for elevating your mood.

- The salmon is loaded with mood-stabilizing omega-3 fatty acids, which play a role in mental well-being by raising levels of serotonin in the brain.

- The pasta suggested for lunch will help prevent serotonin levels from dipping too low.

- The tomato-based marinara sauce is rich in selenium, a mineral needed in the diet to prevent depression. Other depression-defeating nutrients include vitamin C in the grapefruit, folate in the spinach and avocado, vitamin B_{12} in the fish and dairy products, niacin in vegetables and dairy products, and calcium and vitamin D in the dairy products. Brown rice is high in thiamin, a B vitamin that helps reduce mood-sapping fatigue.

- The protein-rich foods in this menu—fish, dairy products, brown rice, and red beans—supply various types of good-mood amino acids. The fish, beans, and oats (in the oatmeal and granola bar), for example, are high in tryptophan, a building block of serotonin. These same foods are also rich in tyrosine and phenylalanine, two amino acids that ward off bad moods.

- The frozen yogurt or sherbet in the bedtime snack is high in glucose and protein, which will enhance the production of mood-lifting serotonin.

Memory: When a Simple Sugar (Glucose) Turns Good

Feeling like the absentminded professor lately? Fortunately, you can bring back a faltering memory by populating your diet with memory-boosting foods, and one of the most surprising is the simple sugar, glucose. Through studies with animals and humans, scientists have discovered that a jolt of glucose, taken after fasting or following a meal, will improve long-term memory, the part of your memory that stores most of everything you know.

Case in point: In a University of Virginia study, investigators served a group of college students lemonade containing 50 grams of glucose (about 200 calories' worth of sugar), then gave them a number of cognitive tests. Drinking the glucose-spiked lemonade greatly enhanced the students' mental performance, particularly on a reading retention exercise. Other fascinating research has found that people with Alzheimer's disease, Down's syndrome, and head injuries perform much better on mental tests when given glucose. Scientists do not know exactly why or how glucose enhances memory so well, but investigations into this benefit are ongoing. One good source of liquid glucose is a sports drink such as Gatorade. Used to replace lost fluid after intense exercise or strenuous work, sports drinks are a mixture of water, carbohydrate (mostly glucose), and minerals called electrolytes.

Glucose as a memory enhancer certainly looks promising, and if used for this purpose, it could be classified as a "good carb." But before you rush out and get some sugary lemonade or glucose-containing sports drinks, let me caution you that to date no one knows what the optimum dose is for mental performance. You might be able to chug 50 grams of glucose in a beverage with no problems at all, but someone else might get high blood sugar due to metabolic differences. So do some experimenting to see how much is right for you.

Here's a typical menu to un-muddle your mind and enhance your memory. (Foods marked by an asterisk indicate that the food is a carbohydrate.)

The Remembrance Menu

Breakfast

*½ cup cream of wheat, sprinkled with 2 tablespoons *wheat bran

2 eggs, any style

*½ cup blueberries

1 cup green tea, iced or hot

Mid-Morning Snack

*8 ounces tomato juice blended with 1 tablespoon brewer's yeast

3 tablespoons almonds or walnuts

Lunch

Tuna salad on a bed of *green leafy lettuce

1 banana

1 cup green tea, iced or hot

Mid-Afternoon Snack

*4 whole-wheat crackers

2 ounces sliced turkey

*1 cup glucose-containing sports drink

Dinner

12 baked oysters

*½ cup brown rice

*1 cup cauliflower

1 cup fruit-flavored yogurt

Why It Works

- Eggs and dairy products are excellent sources of choline for producing acetylcholine, the memory neurotransmitter.

- In recent animal studies, blueberries have been identified as a fruit that enhances short-term memory. The fruit is well endowed with antioxidants and phytochemicals (plant chemicals) that protect the brain against degeneration. Scientists speculate that antioxidants in blueberries reduce inflammation, a process that might harm brain tissue as we get older. Blueberries are among the richest fruit sources of antioxidants. Tomato juice and green leafy vegetables also contain important antioxidants.

- Folate is present in brewer's yeast and green leafy lettuce. Among its many other duties in brain health, folate helps prevent dementia.

- Brewer's yeast and brown rice are a top sources of thiamin (vitamin B_1), which is required for the synthesis of acetylcholine and is involved in improving learning capacity.

- Tuna, turkey, whole wheat, eggs, brewer's yeast, bananas, and cauliflower all contain vitamin B_6, which helps boost long-term memory.

- Tuna is high in docosahexaenoic acid, or DHA, a key memory-enhancing fat. Because of its importance in human brain tissue, DHA might help prevent degenerative brain diseases such as dementia, memory loss, and Alzheimer's disease.

- Oysters, tuna, and wheat bran are loaded with zinc, a mineral that plays a vital role in memory formation and retention.

- The fish, poultry, and brewer's yeast in this menu provide the B vitamin niacin, which dilates blood vessels so more blood, oxygen, and nutrients can reach and nourish your brain.

- Nuts such as almonds or walnuts are good sources of vitamin E, which appears to protect against degenerative brain diseases such as Alzheimer's disease.

- Drinking green tea is recommended for memory preservation. It protects against cognitive decline in the elderly by reducing bodily levels of cholesterol and homocysteine, a harmful protein that has been implicated in heart disease. Both substances are associated with elevated amounts of beta amyloid peptides, proteins that form plaques in the brain and lead to Alzheimer's disease (see below).

Brain Protection with Good Carbs

Unfortunately, your brain is vulnerable to various illnesses, some of them serious, debilitating, and life-threatening. One of the scariest is dementia. Dementia is a condition in which you gradually lose the ability to remember, think, reason, interact socially, and care for yourself. It is not a disease, but rather a cluster of symptoms triggered by diseases or conditions that adversely affect the brain. Some of these triggers can be treated and are referred to as "reversible" dementia; others cannot be cured and are termed "irreversible" dementia. Examples of irreversible dementia include Alzheimer's disease and multi-infarct dementia.

Reversible Dementia

Two forms of reversible dementia are nutrition related. "Nutritional" dementia is one of these. Virtually any shortfall of nutrients, particularly the B vitamins, will cause dementia. Among other duties, B vitamins are an important factor in regulating the health of your nerves, so it should come as no surprise that they are intimately involved in the workings of the brain. B vitamins are most plentiful in whole grains, a group of good carbs. But if your diet is too high in refined foods with lots of added sugar, you could be putting yourself at risk of a B vitamin deficiency, because sugar destroys these important brain-protective nutrients.

Another form of reversible dementia has to do with alcohol dependence. Alcohol, a bad carb, is essentially a toxin that, if abused, can damage your brain, cause memory loss, and induce dementia. In fact, chronic alcoholism can cause a type of amnesia in which the brain is unable to form new memories. Treatment for alcohol dependence and the dementia it causes is very straightforward: abstinence.

Alzheimer's Disease

Among the most feared of all brain diseases is Alzheimer's disease. If it strikes, you gradually lose your mind, your memory, and the ability to recognize your loved ones. In the advanced stages of the disease, you become totally dependent on others for your care. Alzheimer's disease is unusual in that it can be definitively diagnosed only after death through an autopsy of the brain, although mental functioning tests can be administered to detect the possibility of the disease.

Happily though, huge strides have been made in learning how to delay and even prevent the symptoms of this horrifying disease, as well as how to treat it effectively. One of the best ways to possibly escape this terrifying disease appears to be through healthy living, pure and simple.

Part of this lifestyle strategy involves good nutrition, specifically eating a diet high in the B vitamin folate. This nutrient battles the effects of homocysteine in the body, a protein that contributes to clogged arteries and heart trouble. Lots of folate in the diet is believed to guard brain cells from damage by homocysteine—so eat plenty of oranges, orange juice, fortified cereals, and green leafy vegetables. For extra nutritional insurance, take in at least 400 micrograms of folic acid a day, the amount found in most multivitamins.

Multi-Infarct Dementia and Stroke

This common form of dementia is caused by a series of strokes (bleeding or lack of blood supply in the brain) that leave pockets of dead brain cells (infarcts). The accumulated effect of these strokes leads to gradual loss of memory; personality changes; depression; sudden, involuntary laughing or crying; partial paralysis of one side of the body; and other symptoms. Another term for multi-infarct dementia is "vascular dementia." It can coexist with Alzheimer's disease.

Although irreversible, multi-infarct dementia is largely preventable by taking measures to reduce your risk of stroke, as well as your risk of high blood pressure and atherosclerosis (the narrowing and thickening of arteries), two conditions that can lead to stroke. Anti-stroke measures include controlling your weight, cholesterol, and salt intake; getting regular exercise; quitting smoking; avoiding or decreasing the frequency of situations known to cause stress in your life; and getting regular medical checkups.

Nutritionally, there is a super-strong connection between diet and stroke prevention. The very best protection against stroke is to eat a diet high in fruits and vegetables. A Harvard study revealed that drinking citrus juices and eating cruciferous vegetables such as broccoli, cauliflower, and Brussels sprouts can reduce your stroke risk by as much as 32 percent. Orange juice reduced risk of a blood clot stroke by 25 percent, says another study.

Why do fruits and vegetables offer such amazing protection? For one thing, they are rich in antioxidants such as vitamin C, found in citrus fruits and vegetables, and beta-carotene and other carotenoids, plentiful in yellow, orange, and red vegetables. Your brain consumes more oxygen than any other organ in your body. Thus, it is highly vulnerable to oxidation, a tissue-damaging process that occurs when oxygen reacts with fat. The by-products of this reaction are disease-causing free radicals.

Fortunately though, oxidation and the free radicals it produces can be neutralized by antioxidants, available from carbohydrates and other foods, supplements, and found naturally in your body.

The best way to increase your supply of brain-protective antioxidants is to eat plenty of fruits and vegetables every day—at least five servings of vegetables and three servings of fruits daily. Include a variety of green, orange, yellow, and purple fruits and vegetables in your diet.

It bears repeating that fruits and vegetables are loaded with folate, which has amazing stroke prevention powers because it prevents the build-up of homocysteine in your body. Homocysteine causes the cells lining arterial walls to deteriorate in three ways. It damages the walls of blood vessels, causing them to constrict; it triggers abnormal blood clotting; and it promotes the build-up of plaque. All these factors conspire to increase your risk of stroke. (High homocysteine levels also deflate your mood and cripple your mental acuity.) The "cure" for elevated homocysteine levels is as easy as eating more folate-rich foods like green leafy vegetables and oranges, drinking orange juice, plus taking supplemental folic acid (400 micrograms a day).

Another stroke-preventive nutrient found in fruits and vegetables is the mineral potassium. Not getting enough potassium in your diet can increase your chance of stroke by 50 percent, according to research. Potassium is plentiful in carbs such as bananas, potatoes, California avocadoes, lima beans, tomato juice, spinach, and orange juice.

For added protection, be sure to eat carbs high in magnesium, a mineral that is widely distributed in foods. Swiss researchers discovered in 1999 that foods rich in magnesium were highly protective against stroke. Magnesium helps regulate blood pressure, maintain normal function of nerves, and protects blood vessels from damage. Carbs such as green vegetables, wheat germ, soybeans, figs, corn, and apples are packed with magnesium.

Clearly, a lot of what goes wrong with your brain, though distressing, is eminently preventable or highly treatable—mostly by adopting healthier lifestyle habits. Those habits, including a diet filled with good carbs, can go a long way toward outsmarting problems with your mind, your mood, and your memory.

• F I V E •

The Carb-Cancer Connection

If there were ever a case to be made for avoiding bad carbs, their link to cancer would seal it shut. Highly refined carbohydrates such as sugar, white flour, and other bad carbs are tied to an increased risk of at least four types of cancer: breast, colon, lung, and pancreatic. Consider the following evidence.

Breast Cancer

Several studies conducted in recent years have looked into the link between bad carbs and breast cancer—the second leading cause of cancer death in women after lung cancer. In 2001, Italian researchers reported in the *Annals of Oncology* that high-glycemic index foods such as white bread increased a woman's chances of getting breast cancer by 40 percent, while medium-glycemic index foods like pasta did not affect risk. The researchers based their conclusions on their analysis of nearly 2,500 dietary records of women who had breast cancer.

Colon Cancer

Studies of large populations have turned up rather consistent findings: that diets high in added, refined sugar are associated with a greater risk of colon cancer, the third most common form of cancer found in men and women in the United States. In a 1997 study, for example, researchers at the University of Utah analyzed the dietary records of nearly 2,000 people and found that those who ate a lot of sugar and simple carbs were at a greater risk than people whose diets were high in healthier, complex carbs. In a similar study, conducted in 2001, Italian researchers looked into the relationship between high-glycemic index diets and cancers of the colon and rectum by calculating the average daily glycemic index and fiber intake of roughly 2,000 people who had filled out food questionnaires. The researchers discovered that the odds of getting either of these cancers went up considerably with a diet that was high in refined carbohydrates.

Lung Cancer

Lung cancer is the deadliest form of cancer among men and women. More people die annually of this cancer than of colon, breast, and prostate cancers combined. While the greatest lifestyle risk factor for lung cancer is smoking, another might be a high-sugar diet. That's the finding of a study published in *Nutrition and Cancer* in 1998 by a team of researchers from Uruguay. They analyzed the dietary records of 463 people with various types of lung cancer, and after removing smoking and other factors from the research equation, they found that a high intake of sugar was associated with a greater risk of lung cancer.

Pancreatic Cancer

Pancreatic cancer is the fourth leading cause of cancer death in men and women in the United States. Looking into a link between pancreatic cancer and carbohydrates, a team of researchers from the National Cancer Institute analyzed the dietary records of women participating in the Nurses' Health Study, an ongoing study of the relationship between health habits and disease in nearly 90,000 nurses who have been tracked since 1976. Zeroing in on the women's sugar and carbohydrate intakes, the researchers discovered that sedentary, overweight women who ate meals high in added sugar and white flour (both bad carbs) had a 53 percent increased risk of pancreatic cancer.

Understanding the Carb-Cancer Connection

In these cancers, the major connection to carbohydrates hinges on the carbs ranked as "fast" on the glycemic index scale, one of the scorecards used in chapter 2 to rate carbs. Eating fast carbs makes your glucose and insulin levels go sky high. This in turn might raise levels of "insulinlike growth factors." Insulinlike growth factors are chemicals in the body that might promote cancer when produced in excess by increasing abnormal cell growth that leads to cancer.

Soaring insulin levels, which are the result of a diet too high in refined carbohydrates, can also lead to insulin resistance, a metabolic disturbance that is linked to cancer risk. In insulin resistance, cells don't respond to insulin properly. Because insulin's job is to help your cells use and store glucose from the digestion and absorption of food, cells receive no nourishment and, thereby, run the risk of impaired function.

Beyond the insulin connection, bad carbs are practically devoid of cancer-fighting food components such as antioxidants, phytochemicals, fiber, and many other vital food factors. Take refined flour, for example. It gets stripped of 60 to 90 percent of its critical vitamins and minerals during the milling process.

From another angle, several types of bad-carb junk foods, namely snack chips and french fries, contain alarmingly high levels of a cancer-causing chemical called acrylamide, according to tests commissioned by the Center for Science in the Public Interest (CSPI), as well as studies conducted by the Swedish government. Acrylamide forms as a result of chemical reactions that take place during baking or frying at high temperatures. It is found in high amounts in potato chips, corn chips, frozen french fries, and fast-food french fries. In fact, the amount of acrylamide in a large order of fast-food french fries is around 300 times more than what the U.S. Environmental Protection

Agency (EPA) allows in a glass of water! (Acrylamide is sometimes used in water-treatment processes.) Acrylamide is one more reason to avoid bad carbs that contain it.

From a global perspective, the number of new cancer cases is expected to rise by 50 percent over the next 20 years, according to new data from the World Health Organization (WHO). The reason for this increase is because poor nations are adopting bad Western habits, like smoking, drinking, eating the wrong kinds of carbs and other foods, and not exercising. It just so happens that rich nations, like the United States, have higher rates of cancer than poor ones, mostly because of unhealthy lifestyles.

When you consider all the evidence, one thing is clear: If you want to live a lifestyle associated with a low risk of cancer, cutting back on sugar, refined carbohydrates, junk food, and high-glycemic carbohydrates is a powerful change you can make right away.

Although the evidence indicting bad carbs is fairly strong, there is even stronger evidence showing that good carbs, namely fruits, vegetables, and whole grains, are powerful protectors against cancer. In fact, good nutrition is now recognized by scientists and medical experts as playing a major role in slashing the risk of many cancers. The American Institute for Cancer Research (AIRC) estimates that 375,000 cases of cancer could be prevented every year in the United States if we made better dietary choices. Plus, Harvard's School of Public Health suggests that improved diets, more exercise, and other healthy lifestyle habits might help cut cancer deaths by as much as one third. Clearly, there is growing evidence that many good carbs, along with a healthy lifestyle, can help you prevent many different types of cancers. For information on which carbs might prevent certain types of cancer, refer to the following chart.

Cancer-Fighting Carbs

Eat more good carbs and you'll become healthier—and perhaps free of cancer. This message is coming through loud and clear from research studies, major medical organizations, physicians, cancer specialists, and many other sources. But to make the right dietary adjustments, you need to know which cancer-fighting nutrients are found in which carbs. Here is a rundown.

CANCERS	POTENTIALLY PROTECTIVE GOOD CARBS
Bladder	Garlic, green leafy vegetables, soy foods, yellow/orange vegetables, and yogurt and other fermented milk products
Breast	Apples, bran, beans and legumes, broccoli, Brussels sprouts, button and shiitake mushrooms, cabbage, carrots and carrot juice, cherries, garlic, kohlrabi, low-fat dairy products (with the exception of skim milk), radishes, soy foods, spinach, whole grains, yellow/orange vegetables, and yogurt
Cervical	Romaine lettuce and other green leafy vegetables, tomatoes and tomato products, and yellow/orange vegetables
Colon	Broccoli, Brussels sprouts, cabbage, carrots, cauliflower, celery, garlic, grapes and grape juice, kale, legumes, lettuce, low-fat dairy products, oat bran, oranges and orange juice, spinach, tomatoes and tomato products, wheat bran, whole grains, and yogurt and other fermented milk products
Endometrial	Broccoli, Brussels sprouts, cabbage, cauliflower, kale and other green leafy vegetables, and yellow/orange vegetables
Esophageal	Tomatoes and tomato products
Liver	Garlic
Lung	Broccoli, Brussels sprouts, cabbage, carrots and other yellow/orange vegetables, cauliflower, hot peppers, kale, onions, oranges, spinach and other green leafy vegetables, and tomatoes and tomato products
Oral	Tomatoes and tomato products

(continued)

CANCERS	POTENTIALLY PROTECTIVE GOOD CARBS
Ovarian	Broccoli, Brussels sprouts, cabbage, cauliflower, kale and other green leafy vegetables, and yellow/orange vegetables
Pancreatic	Legumes and tomatoes and tomato products
Prostate	Brussels sprouts, broccoli, cabbage, canola oil, cauliflower, kale, low-fat dairy products, soy foods, and tomatoes and tomato products
Stomach	Broccoli, Brussels sprouts, cabbage, cauliflower, fava beans, garlic, green tea, kale, onions, oranges and other citrus fruits, tomatoes and tomato products, and whole grains

Antioxidants

As you will recall, antioxidants are vitamins and minerals that protect the body against disease-causing free radicals. To foil these cellular terrorists, antioxidants step in, scour them from the body, and prevent new ones from being formed. Where cancer is concerned, three of the main antioxidants are vitamin C, vitamin E, and selenium—all plentiful in good carbs such as fruits, vegetables, and whole grains. Here's a closer look.

Vitamin C

There's plenty of evidence linking vitamin C to the prevention of cancer. More than a dozen studies have shown that vitamin C, a powerful antioxidant, reduces the risk of almost all forms of cancer, including cancers of the bladder, breast, cervix, colon, esophagus, larynx, lung, mouth, prostate, pancreas, and stomach. Best news of all: Most of the evidence for this anticancer benefit comes from studies of high vitamin C intake from foods, not from supplements!

As a cancer fighter, vitamin C appears to work in a couple important ways. It disarms free radicals before they can damage DNA (which controls cell growth and reproduction) and stimulate tumor growth. It assists the body's own free-radical defense mechanism by working in partnership with vitamin E to arrest free radicals.

The best sources of vitamin C in the diet are good carbs such as citrus fruits. Other foods, such as green and red peppers, collard greens, broccoli, Brussels sprouts, cabbage, spinach, cantaloupe, and strawberries are also excellent sources, and you should try to eat a variety of these foods. In addition, there are other ways to get more vitamin C in your diet:

- Eat at least one citrus fruit daily. Oranges are an exceptional choice because they contain flavonoids (see below) that enhance the absorption of vitamin C.

- Take in extra vitamin C by drinking citrus juice rather than sodas.

- Eat fresh, raw sources of vitamin C whenever possible. Cooking destroys much of the vitamin C in foods.

Vitamin E

Vitamin E is emerging as an important nutrient for fighting one type of cancer: colon. In a five-year study of more than 35,000 women, researchers observed that those who developed colon cancer were the same ones who had low dietary intakes of vitamin E, according to food questionnaires that were filled out and analyzed. The researchers speculated that vitamin E protected cell membranes in the colon against destruction that could lead to cancer.

Vitamin E also inhibits the formation of carcinogenic substances known as "nitrosamines," which are formed from chemicals called nitrates and nitrites, sometimes found in processed meats like hot dogs and salami. Nitrosamines have been implicated in the development of stomach cancer.

Vitamin E is a fat-soluble vitamin, meaning that it can be stored with fat in the liver and other tissues. Vitamin E is also a component of cells, sandwiched between the fatty layers that make up cell membranes. When free radicals come along, they hitch up to vitamin E, damaging it instead of the rest of the cell membrane. In the process, vitamin E soaks up the free radicals, and the cell is protected from damage. Of all antioxidant nutrients, vitamin E does the best job of scavenging free radicals.

Vitamin E is found widely in foods, particularly vegetable oils. One important carb source of vitamin E is wheat germ. Fruits and vegetables also supply appreciable amounts. To increase your vitamin E intake from natural carb sources:

- Sprinkle wheat germ on your cereal in the morning, or mix wheat germ with yogurt.

- Eat an apple a day (apples contain vitamin E).

- Use dried black currants in lieu of raisins, because currants contain vitamin E.

- Stick to vitamin E–rich whole grains such as wheat, rice, oats, rye, and barley, rather than refined grain products, because vitamin E is removed in the milling process.

Selenium

Among the chief antioxidant minerals is selenium. It works by producing glutathione peroxidase, an antioxidant enzyme that can turn troublesome free radicals into harmless water. This mineral also works closely with vitamin E in protecting the body against free radicals.

The mineral might be an important safeguard against cancer. Studies have found that people with high levels of selenium have lower rates of skin cancer, prostate cancer, colon cancer, and lung cancer.

Selenium is plentiful in carbs such as whole grains and legumes. To get more selenium-rich carbs in your diet:

- Go meatless several times a week by serving bean- and legume-based dishes.

- Opt for whole grains over refined cereals. (Selenium is another nutrient removed during the milling process.)

- Select high-selenium foods. Some of the best vegetable sources include onions and broccoli.

- Cook with garlic, another good source of selenium.

Phytoestrogens

Tofu and other soy-based carbs are loaded with natural, hormonally active compounds called phytoestrogens. These substances attach themselves to cancer cells and prevent real estrogen from entering the cells and allowing cancer to grow. Phytoestrogens also act as antioxidants that have been shown to fight free-radical damage that can lead to cancer.

Evidence of the power of phytoestrogens can be seen among people who eat a lot of soy foods. Asian women, for example, eat low-fat diets with large amounts of tofu and other soy-based products. They have five times less the rate of breast cancer than women who eat a typical Western diet. When breast cancer does strike Asian women, it takes a more favorable course and has a higher cure rate.

Men shouldn't feel left out. Because of their regulating action on hormones, phytoestrogens might help prevent prostate cancer, another hormone-dependent cancer. Scientists also feel that phytoestrogens in soy foods might directly inhibit the growth and spread of hormone-needy cancer cells. In fact, soy consumption is proving to be more protective against prostate cancer than any other dietary factor, according to a prostate cancer mortality study conducted in forty-two countries. In test tubes, soy protein actually kills off prostate cancer cells.

A good move is to include more soy products in your diet, particularly as substitutes for meat or milk in low-fat cooking. Here are some suggestions for sneaking more phytoestrogens into your diet:

- Select foods that are highest in phytoestrogens. Some of the best sources are soybeans and soybean products.

- Use soy milk on your cereal and blend into smoothies.

- Try soy burgers in place of hamburgers.

- Use textured soy protein in recipes calling for ground beef.

- Snack on soy-based nutrition bars rather than on candy bars.

- Use tofu on crackers and rice cakes and in Italian recipes like lasagna to replace all or part of the ricotta cheese. Tofu can also be blended into shakes and smoothies, plus used as a base for dips.

- Munch on soy nuts or soy chips, available at most health food stores.

- Replace some of the flour in recipes with soy flour.

Flavonoids

Packing a wallop of power against cancer are a group of natural substances called flavonoids, found abundantly in fruits, vegetables, and whole grains. More than 5,000 flavonoids have been discovered in nature, and many are responsible for the bright colors of the fruits and vegetables you eat. Flavonoids are also essential for the proper absorption of vitamin C, one of the most important antioxidants and cancer fighters. In fact, flavonoids are what make natural vitamin C (found in foods) more effective than synthetic supplemental vitamin C by improving and prolonging the function of the vitamin. Flavonoids and other beneficial plant compounds might be the reason why women who eat lots of fruits, vegetables, and whole grains have far fewer cases of breast cancer.

Some of the more familiar flavonoids are catechins, citrin, hesperidin, quercetin, and rutin. Of these, catechins and quercetin have been the best studied for their protection against cancer.

Catechins are abundant in green and black teas, red wine, and chocolate. Research with animals shows that catechins inhibit cancers of most major organs: skin, lung, esophagus, stomach, liver, small intestine, colon, pancreas, bladder, and mammary gland.

Quercetin is among the top flavonoids in our diets, present in fruits, vegetables, and tea. As noted earlier, research shows that quercetin is protective against prostate cancer, lung cancer, and skin cancer.

Scientists believe that quercetin, catechins, and other flavonoids exert their anticancer effect in three possible ways. First, they act as antioxidants, squelching free radicals. Second, flavonoids appear to interfere with the growth and spread of tumors, possibly by inhibiting "tumor angiogenesis." This is an abnormal process by which new blood vessels are formed to feed tumors. When these supply routes are cut off, tumors can't get the oxygen and nutrients they need to grow. And third, flavonoids fight cancer by increasing the ability of cells to flush out carcinogens.

All fruits and vegetables are well endowed with a variety of cancer-fighting flavonoids. Here are some other ways to increase flavonoid-rich carbs into your diet:

- Season your foods with chopped garlic or onion.

- Eat a salad every day.

- Eat a variety of fruits and vegetables.

- Try a new fruit or vegetable every week.

- Include at least two vegetables with lunch and dinner.

- Double your portion of vegetables at lunch or dinner.

- Top your breakfast cereal with fresh berries.

- Add extra vegetables to soups and stews.

- Go meatless several times a week.

- Eat vegetable burgers, rather than hamburgers, more frequently.

Carotenoids

Don't risk cancer by shortchanging yourself on orange, red, and yellow fruits and vegetables. As mentioned previously, these good carbs are loaded with carotenoids. Beta-carotene, the best known of the carotenoids, might reduce the risk of cancers of the colon, rectum, breast, uterus, and prostate.

Other important carotenoids include alpha-carotene, shown to protect against breast cancer; beta-cryptoxanthin, which looks promising against breast cancer and lung cancer; lutein, which might guard against cancer of the colon, lung, and breast; lycopene, which appears to be protective against cancer of the prostate, colon, bladder, and pancreas; and zeaxanthin, which is linked to a lower risk of breast, cervical, and colon cancers.

Here are some tips for super-charging your diet with carotenoids:

- Fill your plate with as many colorful vegetables as you can. The more colorful your food selections, the more carotenoids you'll eat.

- Eat canned soups with a tomato base.

- Add a jar or two of strained carrots (yes—baby food!) to soups or stews; it is loaded with carotenoids.

- Drink vegetable juices rather than sodas.

- Eat a hefty serving of tomatoes or tomato-based foods at least twice a week or more.

- Add extra tomato sauce or paste to soups or stews.

- Eat sandwiches and salads with tomatoes.

- Snack on raw fruits and vegetables to get the most carotenoids. One exception, though, is carrots, which actually release more carotenoids when cooked.

- Enjoy exotic fruits such as guavas or mangoes for a change of pace.

- Blend cooked carrots or pumpkin into a smoothie.

Folate

Folate, a member of the B-complex family of vitamins, is critical for the synthesis, repair, and functioning of DNA, the genetic material of cells. In fact, numerous scientific experiments have revealed that folate deficiencies cause DNA damage that resembles the DNA damage in cancer cells. This finding has led scientists to suggest that cancer could be initiated by DNA damage caused by a deficiency in this B-complex vitamin.

Research shows that folate suppresses cell growth in colon cancer. It also prevents the formation of precancerous lesions that could lead to cervical cancer—a discovery that might explain why women who do not eat many vegetables and fruits (good sources of folate) have higher rates of this form of cancer. Other studies link low intake of folate to an increased risk of breast, lung, uterine, and pancreatic cancers.

Folate is found in a wide variety of foods, mainly vegetables, cereals, and grains. To get more of this amazing nutrient in your diet:

- Prepare salads using the darkest-leaf lettuce possible (such as romaine). These varieties are higher in folate.

- Incorporate spinach and other green leafy vegetables into recipes such as those for soups, lasagna, and casseroles.

- Drink a glass of orange juice most days of the week. It's loaded with folate.

- Switch to folate-fortified cereals.

Fiber

Every time you bite into a juicy apple, you're eating fiber, an indigestible but indispensable part of food. An ever-growing body of research suggests that you can reduce your odds of getting cancer by increasing your fiber intake to the recommend level of 25 to 35 grams a day. Studies have found that eating more fiber:

- Might reduce the risk of colon cancer by up to 40 percent.

- Lowers the risk of stomach cancer by 60 percent, especially if that fiber is cereal fiber.

- Cuts your risk of cancers of the mouth and throat by half.

How exactly does fiber fight cancer? It does so by detoxifying and eliminating harmful dietary factors and carcinogens from your body. One type of carcinogen eradicated from the body by fiber is bile acid, a by-product of fat digestion. Increased levels of bile acid in the stool can cause cells in the mucous membranes of the intestines to grow abnormally and possibly lead to colon cancer.

Fiber also eliminates hormonal by-products (including estrogen) made by your body that can lead to breast cancer. Normally, these hormonal by-products are excreted in your bowel, but if you shortchange yourself with too little fiber in your diet, food will take days to travel from entry to exit, creating chronic constipation. With constipation, these by-products remain in your system and are reabsorbed back into your body. So by preventing constipation, fiber might indirectly prevent breast cancer. But not only breast cancer: Regularity is important in preventing stomach, pancreatic, and prostate cancer. So making sure there's plenty of fiber in your diet is an excellent way to prevent constipation and, thus, keep cancer-causing by-products moving naturally out of your body.

The key is to populate your diet with more high-fiber foods. Some of your best fiber bets include vegetables, fruits, cereals, and whole grains. To pump up your fiber intake:

- Eat a wide variety of natural foods in reasonable amounts. (There are at least six different types of fiber in foods, and all are vital to good health. By varying your food selections, you ensure that you also eat a variety of these important fibers.)

- Eat a large salad most days of the week. Choose darker-leaf lettuces; they're higher in fiber and other nutrients.

- Eat fruits and vegetables raw and unpeeled whenever possible. (They have far more fiber than foods that have been peeled, cooked, and otherwise processed.)

- Choose whole-grain cereals such as oatmeal, oat bran, bulgur wheat, and brown rice throughout the week.

- Select dry, unsweetened cereals to which extra fiber has been purposely added as part of the product formulation.

- Add high-fiber grains such as barley or brown rice to vegetable soups.

- Sprinkle raw wheat bran over cereals and salads. In research, bran has been shown to reduce levels of hormonal by-products and other potential carcinogens.

- Substitute high-fiber foods such as beans and lentils for meat and poultry several times a week.

Good carbs, the ones that have been listed and discussed so far in this book, are loaded with protective factors that pay big dividends when it comes to fighting cancer. Variety is the key. When you eat a wide range of fruits, vegetables, and whole grains, you are supplying your body with the full armor of protection against this deadly disease.

The Diabetes Defense

For preventing and treating diabetes and other blood sugar disorders, carbohydrates are the most important food group because they have the most pronounced effect on your blood sugar levels. After you eat carbohydrates, your blood sugar rises, and insulin is secreted in response. Insulin's job is to keep your blood sugar within normal ranges—neither too high nor too low—by shuttling that blood sugar into your cells for energy, thereby lowering levels of glucose in your blood.

Sometimes though, normal insulin activity is upset, and cells cannot use glucose properly. This results in either type I diabetes or type II diabetes, two of the most common blood-sugar-related medical problems. Though treatable, diabetes is the seventh leading cause of death in the United States, and 16 million people have it.

In type I diabetes, the body does not produce any insulin, and consequently, cells cannot absorb glucose. Type I diabetic patients, who typically get the disease at a young age, must, therefore, rely on daily injections of insulin to survive. You're at a greater risk of developing type I diabetes if your siblings or parents have the disease.

Type II diabetes is the most common form of the disease, accounting for 90 to 95 percent of all cases. In type II diabetes, the body can't make enough

WARNING SIGNS OF DIABETES

Type I Diabetes	*Type II Diabetes
• Frequent urination	• Any of the type I symptoms
• Increased thirst	• Frequent infections
• Increased appetite	• Blurred vision
• Unexplained weight loss	• Slow-healing wounds
• Extreme fatigue	• Tingling or numbness in hands or feet
• Irritability	• Recurring skin, gum, or bladder infections

*Often, people with type II diabetes experience no symptoms.
Source: American Diabetes Association

insulin or use it properly. Consequently, glucose does not get into cells as it should, and there is a glut of glucose in the blood.

Risk factors for type II diabetes are linked largely to age and lifestyle factors, although genetic factors are involved, too. With age, for example, your body gradually loses its ability to regulate glucose. This puts you in danger of developing the disease, which can typically occur in people over 45 years old, particularly among those who are overweight. In fact, more than 80 percent of type II diabetics are overweight. What's more, type II diabetes is present in 25 percent of the population age 85 and older.

Diabetes has been dubbed the "silent killer" because most people are unaware that they have it until diagnosed with one of its life-threatening complications, such as blindness, kidney disease, nerve disease, or cardiovascular disease. In fact, the American Diabetes Association estimates that there are 5.4 million people in the United States who have diabetes but don't know it. There are some warning signs, however, and these are listed above.

Diet and Diabetes

The primary treatment for type I diabetes is insulin, taken by injection. Type II diabetes, on the other hand, can be managed by changes in diet and exercise and often by drugs other than insulin.

If you have either form of diabetes, food is one of the chief tools you can use to control the disease and stay healthy. Food helps keep your blood sugar in line, provides energy for exercise and daily activities, and supplies the nutrients needed for health and healing. Food is nourishment and medicine all rolled into one.

What kind of food should you eat if you have diabetes? A prevailing myth about diabetes is that you must follow a special "diabetic diet." Not so. You have the same nutritional needs as anyone else. Like the rest of the world, you need regular, well-balanced meals that provide a variety of nutrients from carbohydrates, protein, and fat. Here is an overview of what you need if you have diabetes:

Carbohydrates

Roughly half of your diet (40 to 50 percent of your total daily calories) should be made up of carbohydrates, with an emphasis on natural, high-fiber choices, such as vegetables, legumes, fruits, and whole grains. Because carbohydrates are so important in diabetes, they will be discussed in detail later in the chapter.

Protein

Protein intake should make up 10 to 20 percent of your total daily calories. Choose lean, low-fat proteins from poultry, fish, dairy, and vegetable sources such as beans, legumes, and tofu. (If you have kidney disease—often a complication of diabetes—you might be placed on a protein-restricted diet, because excess protein places an undue burden on the kidneys. Your doctor will probably recommend that you reduce your protein intake to around 10 percent of your daily calories.)

Fat

Generally, a healthy intake of fat is 30 percent or less of your total daily calories. Less than 10 percent should come from saturated fats (fats that are solid at room temperature). The remaining 10 to 15 percent should come from polyunsaturated fats (available from fish and vegetable oils) and monounsaturated fats (found in olive oil and nuts). Dietary cholesterol should be limited to less than 300 milligrams daily to discourage cardiovascular disease, which is the major complication of diabetes. Cholesterol is found in eggs, meats, and high-fat dairy products.

Good Carbs and Diabetes

Ideally, in a diet to control diabetes you should use all three carbohydrate scorecards to select only the healthiest carbs for your diet. For example:

Simple versus complex

Most of the carbohydrates in your diet should come from the good carbs such as complex carbohydrates. These carbs furnish more vitamins, minerals, antioxidants, and phytochemicals than simple carbohydrates do.

Sugar, a simple carbohydrate, is no longer a forbidden food in diabetes. It is perfectly fine to include a little bit of sugar and sugar-containing foods in your diet, and you don't really have to avoid table sugar, corn sweeteners, syrups, or other such foods. Eating sugar and sugar-containing foods does not hurt blood-sugar control in either type I or type II diabetes. However, you should not go overboard, because sugar is a poor nutritional choice and is tied to heart disease, obesity, and cancer. If you wish to restrict sugar but have a bit of a sweet tooth, opt for artificial sweeteners instead.

Fast versus slow (glycemic index)

You can use the glycemic index to select carbohydrates that are less likely to cause roller-coaster swings in your blood sugar. There are benefits to doing so. Research has found that following low-glycemic index diets, on average, can lower your blood glucose and insulin by as much as 30 percent. Because low-GI foods have appetite-control and fat-burning benefits, incorporating them into your diet can help you shed pounds.

Low fiber versus high fiber

The carbohydrates you choose should be rich in fiber because of its beneficial effects on glucose and insulin metabolism. High-fiber foods require prolonged breakdown and, thus, release blood sugar more slowly. This action helps prevent dips in blood sugar and helps maintain even energy levels throughout the day. A study published recently in the *New England Journal of Medicine* reported that people with type II diabetes could lower their blood sugar and insulin levels by as much as 10 percent simply by eating a fiber-rich diet.

The type of fiber most responsible for this glucose-lowering effect is soluble fiber, found in oat bran, oatmeal, barley, peas, beans, carrots, apples, and oranges.

Try to shoot for 25 to 35 grams of fiber a day from fruits, vegetables, beans, and whole grains.

The chart on the next page lists good carbs that should be included in your diet most days of the week.

Working with Carbohydrates in Your Diet

If you have diabetes, your meal planning should focus on low-fat nutrition, moderate amounts of protein, and designated quantities of high-fiber, complex carbohydrates, with priority given to the amount of carbohydrates you eat. Your diet should also be consistent; that is, you should eat roughly the same amount of calories each day, eat meals and snacks at the same time every day, and not miss meals. In addition, structure your meals to include a variety of many different types of foods.

A meal planning system growing in favor for managing diabetes is "carbohydrate counting." This is a method of counting the grams of carbohydrates you eat at meals and snacks. The reason for counting carbohydrates is because they have the most dramatic effect on your glucose levels. Your body converts carbohydrates into glucose faster than it does protein or fat. In fact, carbohydrate is converted to glucose within the first two hours after you eat a meal.

If you know how much carbohydrate you've eaten, you can predict what your blood glucose will do. A little bit of carbohydrates will elevate your glucose, and large amounts will make it go up even higher. For example, a full cup of cereal will make your blood glucose go higher than a half-cup, and two cups will elevate it even more.

These metabolic facts of life have particular importance if you take insulin, which is required to balance the glucose. It is the amount of carbohydrates in a meal that largely determines how much insulin you require. Counting carbohydrates, therefore, can help you make appropriate insulin adjustments based on your blood glucose patterns.

With this method of meal planning, carbohydrates are computed in grams and usually translated into "carbohydrate choices." For example, 1 carbohydrate choice = 15 grams of carbohydrate. You can find information

TOP 15 ANTIDIABETIC CARBS

Carbohydrate	Beneficial Action
Apples	High in the soluble fiber pectin, which helps control blood sugar; also high in beneficial plant compounds called flavonoids, which help prevent diabetic complications.
Barley, pearled	High in the soluble fiber pectin, which helps control blood sugar; also a low-GI food.
Broccoli	Contains glutathione, an antioxidant that helps reduce the risk of diabetes and its complications.
Brown rice	High in the mineral chromium, which helps regulate blood sugar.
Carrots	Although higher on the glycemic index, carrots contain beneficial substances called carotenoids (beta-carotene is a carotenoid) that might protect against abnormal elevations in blood sugar.
Garlic	Contains a beneficial plant compound called allicin, which helps lower blood sugar.
Green leafy vegetables	Rich in antioxidants and phytochemicals, which help prevent diabetic complications.
Jerusalem artichokes	Contain a beneficial substance called inulin, which helps lower blood sugar after a meal, probably by slowing its absorption in the intestine.
Kidney beans	Among beans and legumes, the kidney bean is the highest in soluble fiber, which reduces the rise in blood sugar after a meal, normalizes insulin levels, and helps keep blood sugar levels steady throughout the day.
Oat bran	High in the soluble fiber pectin, which helps control blood sugar.

(continued)

TOP 15 ANTIDIABETIC CARBS

Carbohydrate	Beneficial Action
Onions	High in beneficial plant compounds called flavonoids, which help prevent diabetic complications.
Pumpkin	Contains beneficial substances called carotenoids (beta-carotene is a carotenoid) that might protect against abnormal elevations in blood sugar. Although higher on the glycemic index, pumpkin is rich in fiber.
Rye bread	Enhances insulin secretion in the pancreas, where insulin is produced.
Soybeans	High in fiber; also high in biotin, a B vitamin that enhances the body's use of insulin and is beneficial in the regulation of blood sugar.
Whole grains	High in fiber; also beneficial in helping the body use insulin; lowers the risk of type II diabetes.

on how much carbohydrate is in foods by checking the Nutrition Facts Panel on food labels, or by consulting a carbohydrate gram counter guide.

Carbohydrate counting has been found to be among the most effective tools ever devised for controlling blood sugar and achieving other treatment goals. It works best if you:

- Need to prevent swings in your blood sugar.

- Want to eat the same amount of carbohydrates at given times to help stabilize your levels of blood glucose.

- Are within a reasonable body weight range.

- Have type I diabetes and need to better match your insulin to the foods you eat.

- Are willing to monitor your blood glucose levels before and after meals and keep records of the results.

- Have had an inconsistent carbohydrate intake in the past.

- Have been unsuccessful using other meal-planning systems.

- Have been newly diagnosed with diabetes.

- Have had diabetes for a long time and are willing to try a new approach to meal planning.

- Like to work with numbers.

How to Plan a Carbohydrate Counting Diet

Again, carbohydrates should make up roughly one-half of your required calories. With carbohydrate counting, carbohydrates are divided equally into meals and snacks. Eating the same quantity of carbohydrates at each meal helps stabilize blood glucose levels.

Thus, you build your daily meals and snacks around a fairly consistent number of carbohydrate grams, which translate into carbohydrate choices. In addition to carbohydrate choices, you'll also choose protein foods and fats. An example of a daily diet plan using this system follows.

If you are interested in learning more about carbohydrate counting, work with a registered dietitian who is well versed in the system. Your dietitian will help you master the system; identify patterns of blood sugar levels that are related to your diet, diabetes medications, and physical activity; and teach you how to calculate the amount of insulin you need to control blood sugar when you eat a specific amount of carbohydrate.

At the same time, educate yourself on how many grams of carbohydrates are in the foods you eat. Initially, you'll have to weigh and measure your foods, but with time, you'll be able to eyeball the correct portions.

Using Special Diabetic Supplements

Although it is generally best to get your good carbs directly from food, you can get a little more of a good thing through diabetic supplements, specially formulated to help you control your blood sugar. For example:

CARBOHYDRATE COUNTING: SAMPLE MEAL PLAN

Meals and Snacks	Carbohydrate Grams	Carbohydrate Choices	Foods
Breakfast	60 grams of carbohydrate	4 carbohydrate choices	½ cup oatmeal, or 1 banana (small), or 1 large bran muffin (counts as 2 carbohydrate choices)
	1 ounce of protein	1 ounce of protein	1 hard-boiled egg
	2 fats	2 fats	2 teaspoons butter or margarine
Lunch	60 grams of carbohydrate	4 carbohydrate choices	1 whole-grain hamburger bun (counts as 2 carbohydrate choices), or lettuce and tomato slices, or low-fat yogurt sweetened with aspartame, or 1 cup raspberries
	3 ounces of protein	3 ounces of protein	3-ounce hamburger patty
	1 fat	1 fat	1 teaspoon mayonnaise
Dinner	60 grams of carbohydrate	4 carbohydrate choices	Salad (1 cup greens; or 1 cup chopped raw salad veggies and ½ cup artichoke hearts); or 1 baked potato; or 1 roll; or ½ cup water-packed fruit cocktail

(continued)

CARBOHYDRATE COUNTING: SAMPLE MEAL PLAN

Meals and Snacks	Carbohydrate Grams	Carbohydrate Choices	Foods
Dinner	3 ounces of protein 2 fats	3 ounces of protein 2 fats	3 ounces grilled salmon 2 teaspoons French salad dressing
Snack	30 grams of carbohydrates	2 carbohydrate choices	3 whole-wheat crackers; or 1 cup skim milk

Nutrition: Approximately 1,800 calories; 53 percent of calories from carbohydrates; 21 percent of calories from protein; 26 percent of calories from fat; 25 grams of fiber.

Adapted from: Benedict, M. "Carbohydrate Counting: Tips for Simplifying Diabetes Education." Health Care Food and Nutrition Focus, 16: 6–9, 1999; and Greenwood-Robinson, M. Control Diabetes in Six Easy Steps, 2002.

Diabetic Snack Bars

What they are

Diabetic snack bars are designed to prevent hypoglycemia, a lower-than-normal concentration of glucose in the blood, or reduce hyperglycemia, abnormally high levels of glucose in the blood. These products can be used as convenient snacks or as an occasional meal replacement if you have diabetes.

How they work

Diabetic snack bars contain special carbohydrates that result in better-controlled blood sugar levels. Some products are formulated with uncooked cornstarch, a slowly absorbed complex carbohydrate that provides a sustained source of glucose. Examples of products containing uncooked cornstarch include Extend Bar, Nite Bite, and Gluc-O-Bar. Studies show that bars containing uncooked cornstarch do an excellent job of preventing low blood sugar. In a clinical trial involving the Extend Bar, the product helped ward off low blood sugar for up to nine hours in diabetic patients.

Other diabetic snack bars contain a special kind of carbohydrate called "resistant starch," which is broken down into glucose more slowly than other carbohydrates. In contrast to uncooked cornstarch, which is slowly but fully digested in the small intestine, resistant starch is not completely digested and is, thus, lower in calories. Bars with resistant starch have been proven in research to stabilize blood sugar levels, preventing abnormal and potentially dangerous spikes in glucose. Examples of products with resistant starch are Choice dm (Mead Johnson Nutritionals) and Glucerna (Abbott Laboratories).

How to use them

Diabetic snack bars with uncooked cornstarch are designed to be used as a bedtime snack because they prevent episodes of hypoglycemia that can sometimes occur overnight (a condition called "nocturnal hypoglycemia"). Bars with resistant starch are meant to help prevent elevations in blood sugar during the day.

Safety

Diabetic snack bars are safe when used as directed and as part of an overall nutritional and medical strategy for managing diabetes.

Diabetic Beverages

What they are

Diabetic beverages are a special type of liquid carbohydrate supplement, that like diabetic snack bars, are formulated for people with diabetes. The two leading products, available in pharmacies and grocery stores, are Choice dm beverages and Ensure Glucerna beverages. Both are available in ready-to-drink cans.

How they work

Most of these beverages are rich in slowly digested complex carbohydrates, protein, fiber, contain little or no fat, and are fortified with vitamins and minerals. They provide therapeutic support to help control blood sugar, particularly in conjunction with diabetes drugs, and are geared toward helping reduce the complications of diabetes. Tested clinically, these products help stabilize blood sugar.

How to use them

Diabetic beverages are meant to be used as a snack or as part of a small meal when eaten with fruit or vegetables.

Safety

These products are safe and nutritious when used as directed.

Nutritional Goals in Type I Diabetes and Type II Diabetes

With type I or type II diabetes, strive to keep your blood glucose level as close to normal as possible by paying attention to your diet. Here's a summary of the key nutritional goals for each form of diabetes.

If You Have Type I Diabetes:

- Normalize your blood glucose level through healthy food choices and good meal planning. With tighter glucose control, you'll feel better and lower your odds of getting eye, kidney, or heart disease.

- Integrate insulin therapy into your eating and exercise patterns.

- Focus on eating the same amount of carbohydrates at the same meal each day. Select good carbs for your diet.

- Monitor glucose levels and adjust insulin doses for the amount of food you eat.

- Plan meals and snacks to prevent hypoglycemia. Consider using diabetic snack bars or beverages.

- Through diet, exercise, and other lifestyle changes, keep your blood fats (cholesterol and triglycerides) in normal ranges to lower your risk of heart disease. Monitor these levels with regular medical check-ups.

- Improve your overall health through good nutrition.

If You Have Type II Diabetes:

- Normalize your blood glucose level, cholesterol, and blood pressure through healthy food choices, good meal planning, and regular medical check-ups.

- Lose weight if you need to.

- Plan meals and snacks to prevent hypoglycemia. Consider using diabetic snack bars and beverages as part of your meal planning.

- Improve your overall diet through the selection of good carbs.

Preventing Diabetes

Type I diabetes is primarily a genetic disease, so if it runs in your family, it might be difficult to prevent. Type II diabetes is a different story, however. Although there is a genetic component to type II diabetes, the wonderful news is that you can prevent it with a lifestyle fix that includes mostly diet, exercise, and other nondrug measures. For example:

- **Good carb nutrition.** Follow a diet that focuses on low-fat nutrition, moderate amounts of lean protein, and good carbs. In planning a preventive diet, 40 to 50 percent of your total daily calories should come from good carbs.

- **Weight loss and control.** Obesity is a major risk factor for type II diabetes. If you are overweight, begin a sensible reducing diet (see chapter 9 for guidelines on weight control) that is low in fat, high in fiber, and moderate in good carbs. It is best to restrict all bad carbs as much as possible.

- **Exercise.** Leading an inactive lifestyle contributes to type II diabetes and expedites its progression in two likely ways. First, inactivity leads to obesity, which is the major promoter of type II diabetes. Second, inactivity makes body cells resistant to using insulin and taking in glucose. To overcome these damaging factors, exercise with a program that includes both weight training and aerobics. These activities stimulate normal insulin activity, help your cells take in glucose, and assist you in burning body fat.

- **Quit smoking.** Nicotine harms the ability of your cells to properly respond to and use insulin. Smoking also generates free radicals, which can do irreparable harm to your body.

Many of these same measures can help you prevent, or manage, a condition called "metabolic syndrome" or "syndrome X," which sets the stage for both type II diabetes and cardiovascular disease. The core problem in metabolic syndrome has to do with insulin. If you have metabolic syndrome, your cells defy the action of insulin (insulin resistance), and as a consequence, blood sugar is locked out of cells. Your cells receive little energy or nourishment, and blood sugar piles up in your bloodstream, creating a toxic metabolic environment.

Red flags for this syndrome are knowable through medical testing and include:

- High blood pressure (greater than 130/85; 50 percent of people with high blood pressure have metabolic syndrome)

- Elevated triglycerides (150 or higher)

- Abnormally low HDL cholesterol (less than 50 for women; less than 40 for men)

- Above-normal blood sugar (higher than 110)

- Central obesity, in which your waist measurement is greater than 35 inches if you are a woman and greater than 40 inches if you are a man

Generally, if you have three or more of the above symptoms, your doctor might diagnose you as having metabolic syndrome.

But how does metabolic syndrome develop in the first place? As it turns out, diet is mostly to blame, particularly a diet that's high in bad carbs such as sweets, breads, and processed snack foods. Such foods trigger a rapid spike in blood sugar, and the body responds by pumping out more insulin into the bloodstream to handle the sugar. Over time, insulin levels in your blood remain higher than they should be. This promotes fat storage, elevates blood fats, and raises blood pressure.

Happily though, metabolic syndrome can be prevented and reversed with weight loss, regular exercise, low-fat foods, and a switch from refined carbs to mostly good carbs.

Diabetes might not be curable, but it is manageable and it is preventable. Through proper diet, exercise, and consistent self-care, you can successfully control your diabetes if you have it, and fight it off if you don't. What's more, the sooner you start changing your lifestyle to become more nutritionally and physically fit, the greater your chances of living a full, vigorous, and healthy life.

Digestive Health

If your doctor could write you a prescription for preventing a lot of the digestive ills facing mankind, that prescription would be, in a word, fiber. And you don't have to go to a pharmacy to get it filled, either. Just eat a variety of fruits, vegetables, cereals, whole grains, and other good carbs, and you'll be doing your digestive system a world of good.

Your digestive system is actually a 30-foot long tube, with an entry at your mouth and an exit at your anus. This system is supported by other organs, including the liver and pancreas, which furnish enzymes and other substances required for digestion. There is a lot that can go right or wrong along the way, but eating foods rich in fiber is one important measure you can take to keep this system in peak health. Fiber prevents, or reduces symptoms of, a wide range of digestive problems, from appendicitis to ulcers.

Appendicitis

The appendix is a wormlike structure about three inches long that is attached to your colon (also referred to as the large intestine). Until recently, the appendix was believed to have no known function, described by evolutionists as a remnant from the days when man's theorized ancestors (apes) were plant-eaters. In apes and other plant-eating animals, the appendix aids in the digestion of plant material.

New medical evidence, however, has found that the human appendix is indeed a critical organ—one that is an integral part of the disease-fighting immune system. The appendix contains lymphatic tissue whose task it is to attack disease-causing organisms in the lower end of the digestive tract.

Thus; your appendix might protect you from serious illness. A study conducted at a German medical center in 1979 found a higher incidence of ovarian cancer among women who had had their appendix removed (appendectomy). Other medical research has linked appendectomies to an increased susceptibility to leukemia, Hodgkin's disease, and colon cancer.

Based on this medical evidence, it is vital that you do what you can to prevent "appendicitis," an inflammation and infection of the appendix. This is an extremely serious condition in which the appendix can burst if it isn't surgically removed, spreading infection (peritonitis) throughout the abdomen.

Eating a high-fiber diet has been proposed as a way to prevent appendicitis. This recommendation stems from studies of populations of people around the world who eat high-fiber or low-fiber diets. In countries where the diet is mostly grain- and vegetable-based (high-fiber diets), people rarely get appendicitis. By contrast, people who eat low-fiber diets with lots of refined carbohydrates—the typical Western diet—tend to have a higher rate of this disease. Why this occurs is unknown, but some scientists believe that lack of fiber slows down the movement of stools, causing pressure to build and forcing fecal matter into the appendix. Once trapped in the appendix, this material begins to undergo fermentation by natural microbes—a process that produces irritating by-products that might provoke an infection.

In the 2002 issue of the medical journal *Gut,* physician Dr. J. Black recounted a personal World War II story that lends support to the appendicitis-fiber connection. At the end of the war, Dr. Black was stationed at a hospital in South Burma, where nearly 1,100 Japanese soldiers were encamped while awaiting repatriation. Among these soldiers, there was an alarmingly high rate of appendicitis—about one case every two to three weeks. Even the Japanese medical officers at the camp were surprised by the number of cases, suggesting to Dr. Black that appendicitis was rare among Japanese troops. Dr. Black suspected that the appendicitis was related to the fact that Japanese soldiers were being fed British rations, which had a very low fiber content compared to the traditional Japanese diet.

Similar observations have been made elsewhere. Case in point: When British and Indian troops were stationed in India from 1936 to 1947, appendicitis was four to six times more common in the British soldiers than in the Indian troops, who ate rations that were three times higher in fiber and contained one third the amount of animal protein than the British diet.

Despite the compelling evidence for dietary influences on appendicitis, many medical experts remain skeptical. Cases of appendicitis have actually fallen worldwide since 1950, and improvements in hygiene and better infection control are often cited as possible reasons for the decline.

Even so, in the United States, some 250,000 appendectomies are performed annually. Although there are some valid reasons for the surgery—such as a confirmed case of appendicitis—at least 20 percent of these procedures turn out to be unnecessary. If your surgeon wants to remove your appendix, make sure there is a good reason such as an accurate diagnosis of appendicitis. CT scans and other advanced forms of x-ray technology are helpful in making a precise diagnosis and eliminating unnecessary surgery.

Bottom line: Eating a high-fiber diet is good medicine, whether or not it helps prevent appendicitis. Try to shoot for 25 to 35 grams of fiber daily from a variety of high-fiber foods, and avoid eating refined or processed carbohydrates.

Constipation

One of the most common complaints heard by doctors is constipation. Medically speaking, constipation is the passage of hard stools fewer than three times a week. Other symptoms such as abdominal bloating and discomfort may accompany constipation, too.

Although constipation is not usually serious, it can incite worrisome complications:

- Intestinal obstruction, which can be a medical emergency.

- Hemorrhoids, a cluster of dilated veins in swollen tissue around the anus.

- Hernia, which occurs when an organ protrudes through the wall of the cavity that normally surrounds it.

- Irritable bowel syndrome, a condition characterized by alternating periods of diarrhea and constipation, often accompanied by cramping.

- Colon or rectal cancer. With chronic constipation, fecal matter stays in your system too long. The more slowly this material passes through your colon, the higher your colon's exposure to any carcinogens in the waste. This can stimulate abnormal cell growth in your colon and rectum.

What causes constipation, exactly? The major culprit is not eating enough fiber, pure and simple. Other causes include inactivity and poor water intake. Constipation can also be the result of certain diseases, such as kidney failure, colon or rectal cancer, diverticulosis (see below), thyroid problems, prolonged bed rest, and stress. Certain medications cause constipation, including antacids, antihistamines, aspirin, blood pressure medications, diuretics, iron or calcium supplements, and mood-lifting drugs.

Changing your diet to include more fiber-rich foods such as beans, bran, high-fiber cereals, fruits, vegetables, and whole grains should correct constipation and stop it from occurring in the first place. Some of the best foods for increasing regularity are wheat bran, fruits, and vegetables. Other lifestyle changes such as becoming more active and drinking at least eight glasses of pure water a day will also keep you regular.

It is perfectly normal to get constipated every once in a while, but notify your doctor if your constipation continues more than three weeks, even after increasing your fiber intake, water intake, and exercise level. Chronic constipation should always be evaluated by your physician.

Diverticulosis and Diverticulitis

Diverticulosis develops when small outpouchings called diverticula, ranging from the size of a piece of confetti to a quarter, form in the colon. Between 20 and 50 percent of all people age 50 and over have diverticulosis. And if you live to age 90, you will have developed some diverticulosis.

Most of the time, it is not a problem. But trouble starts when these diverticula become infected and inflamed (the condition known as diverticulitis), often the result of bacteria, undigested food, or feces becoming lodged within them. The consequences can range from a small abscess to a massive infection or even a perforation of the colon wall. In extreme cases, diverticulitis is life-threatening, should an inflamed diverticulum rupture or perforate and spill intestinal material into the abdominal cavity. This can cause peritonitis, an inflammation of the membrane that covers the abdominal organs. Peritonitis is a medical emergency.

Diverticulosis is generally caused by a low-fiber diet. A diet deficient in fiber produces smaller volumes of stool. To move the smaller stool along the colon and out the rectum, the colon narrows itself by contracting down forcefully. This increases pressure, and over time, high pressure weakens the muscular wall of the colon, causing diverticula to form.

Considering that a majority of people never experience any symptoms, how do you know whether you have diverticulosis? Most diverticulosis is accidentally discovered during routine intestinal tract examinations, such as a barium enema or colonoscopy. Some people with diverticulosis, however, do have symptoms, including constipation, cramping, and bloating. If you have been diagnosed with diverticulosis and suddenly experience pain accompanied by fever, you might be having an attack of diverticulitis. Notify your physician immediately. Treatment for diverticulitis usually involves hospitalization, in which you receive intravenous antibiotics and a liquid diet to rest your colon. In extreme cases, surgery to remove a section of the colon might be required.

The best way keep diverticulosis from starting, getting worse, or progressing to diverticulitis is by eating a high-fiber diet. Fiber widens the colon, easing the pressure and reducing the chances that existing diverticula will rupture or become inflamed. Two of the best carbs you can eat for preventing or managing diverticulosis are bran cereal and whole-wheat bread.

Duodenal Ulcers

A duodenal ulcer is a hole or break in the first part of the small intestine known as the duodenum. These ulcers are two to three times more common than those that form in the stomach (gastric ulcer). They develop when there is an imbalance between the amount of natural gastric acid secretions and the resistance to those secretions by the protective intestinal lining in your body. When the normal balance breaks down, the lining is injured, and the result is an ulcer.

A bacteria called *H. pylori* that lives in the mucous membranes lining the digestive tract is the most common cause of duodenal ulcers. About 95 percent of patients with this type of ulcer are infected with *H. pylori.* Also, the regular use of painkillers, mainly nonsteroidal anti-inflammatory drugs (NSAIDs), such as aspirin and ibuprofen, can increase your risk of developing an ulcer by as much as 40 percent. Smoking is another major risk factor.

Symptoms of a duodenal ulcer include heartburn, a burning sensation in the back of the throat, or stomach pain. Bloating and nausea after meals are common complaints, too.

The usual treatment for ulcers is medication to decrease the amount of acid produced or to coat the lining of your duodenum for protection. In addition, many doctors recommend treatment to eliminate *H. pylori.* The usual therapy involves antibiotics and Pepto-Bismol.

But high-fiber nutrition might play a role, too, and is often recommended as a preventive measure. Fiber might promote healing and prevent a recurrence of the ulcer, according to a study conducted at the Harvard School of Public Health. Although the reason for this ulcer-curbing benefit is unclear, researchers speculate that fiber somehow helps keep the duodenal lining intact by balancing the gastric juices and the resistance of that lining to injury.

Another Harvard study found that fiber protects against the formation of duodenal ulcers. In this study, researchers followed 48,000 men over a 6-year period and observed that those who averaged 30 grams of fiber a day cut their risk of duodenal ulcers in half. A higher intake of vitamin A was also linked to fewer ulcers. High-fiber foods such as apricots, peaches, prunes, carrots, and pumpkin are plentiful in vitamin A and, thus, might deliver a one-two punch against duodenal ulcers. The take-home message here is: Eat a high-fiber diet, with plenty of vitamin A-rich foods, to ward off this digestive disease.

Gallstones

Situated beneath the liver is a pear-shape organ called the gallbladder. It serves a useful purpose—storing bile, a greenish fluid that contains bile acids and cholesterol. Required for the breakdown and digestion of fats, bile is secreted by the liver and makes its way into the small intestine via a passageway called the bile duct.

If cholesterol in the bile duct gets too high, gallstones can form in the gallbladder. These are crystal-like structures that might be as small as a grain of sand or as large as a golf ball. Most gallstones are practically pure cholesterol. They can migrate to other parts of the digestive tract, causing severe pain and life-threatening complications. In rare cases, severe inflammation can cause the gallbladder to rupture, which can be fatal.

If your doctor suspects you have gallstones, you'll undergo blood tests for liver enzyme levels. These levels are usually elevated with gallbladder disease. A definitive diagnosis is made through ultrasound.

Treatment for gallstones varies, depending upon the severity of the situation. Sometimes gallstones can be dissolved by taking bile acids in tablet form or broken up through high-frequency sound waves in a procedure known as lithotripsy. Where there is a severe obstruction and pain, your surgeon might elect to remove the entire gallbladder in a procedure called a cholecystecomy.

Although the gallbladder serves a useful function, the body can get along without it. Still, any type of surgery is stressful on your body, particularly your immune system. One study found an elevated risk of colon cancer following gallbladder removal. Fortunately though, the risk disappears after fourteen years. But if facing gallbladder surgery, seek a second opinion.

The best course of action against gallstones is to prevent them from ever forming. Studies show that eating a poor diet—one that is high in fat, processed foods, sugar, and meat and low in fiber and good carbohydrates—increases your odds of getting gallstones. So it is best to follow a diet that is high in fiber and low in fats, especially saturated fats. Olive oil is recommended (no more than two tablespoons a day) as a healthy fat to include in your diet; it helps lower cholesterol levels in the blood and in the gallbladder. What's more, try to replace processed foods and other bad carbs with complex carbohydrates such as whole grains.

Hemorrhoids

Anal and rectal blood vessels form a tight seal that prevents stool from leaking out, but when these vessels become swollen, often the result of straining during a bowel movement, hemorrhoids form. They can often rupture, causing bleeding.

There are two types of hemorrhoids: those that occur internally near the beginning of the anal canal, and those that occur externally, extending just outside the anus. Once hemorrhoids develop in either of these locations, itching, pain, and bleeding can arise. If you see bright red blood on your toilet paper, in the toilet bowl, or on the stool itself, a visit to your doctor is imperative. The blood could be related to other serious problems, including cancer of the colon or rectum.

As with so many other digestive disorders, the best treatment is a high-fiber diet. By reducing constipation, fiber decreases the likelihood of further hemorrhoids developing. Other remedies to relieve the discomfort include the use of over-the-counter suppositories or ointments to lubricate the rectum and anus for smoother passage of stools, warm baths to shrink the swollen blood vessels, and regular exercise.

Irritable Bowel Syndrome

Sometimes called "spastic colon," irritable bowel syndrome (IBS) is actually a collection of troublesome symptoms: constipation, diarrhea, gas, painful bloating, and nausea. In some cases, you have urgent diarrhea first thing in the morning or during and after meals. In others, you experience painful constipation, which alternates with diarrhea. The good news about IBS, though, is that it is not life-threatening. Nor does it mean that you are at risk for more serious digestive problems.

Women with IBS outnumber men three to one, although no one knows why. Chronic stress and depression trigger IBS symptoms, so part of the problem might be in the mind.

But IBS isn't all mental. It's also in the intestines. In IBS, there is an abnormal muscular activity of the intestinal wall, with bowel muscles contracting up to nine times a minute rather than the normal three. This activity stretches the intestinal wall, causing pain.

Although there is no cure for IBS, you can minimize its symptoms through lifestyle changes such as exercise, stress management, and diet. A high-fiber diet will help relieve the constipation and diarrhea; however, you have to select foods carefully. Beans, for example, can cause unpleasant gas. A small amount of the carbs in beans comes from sugars, namely raffinose, stachyose, and verabose. These sugars are the culprits in the intestinal problems associated with beans. The small intestine doesn't have the right enzymes to digest these sugars, so they arrive in the large intestine undigested. Bacteria residing there have a heyday feeding off these sugars and fermenting them. Carbon dioxide, hydrogen, and other gases are given off in the process. You can relieve discomfort caused by undigested sugars by using a product called Beano, an enzyme preparation that does the digesting for you.

Beans aren't the only gas-producing carbs. Others are onions, celery, carrots, raisins, prune juice, and Brussels sprouts. Never suddenly increase the fiber in your diet; start with small doses so your bowel gets used to it.

Your physician might also recommend that you take a bulk former, such as Metamucil, which contains the natural vegetable fiber psyllium. It might also help relieve your constipation and diarrhea.

Prebiotics: A Special Carbohydrate for Digestive Health

Fiber is not the only nondigestible remnant found in carbohydrates that makes its way through your system. Others are "prebiotics," a special type of undigested carbohydrate that help promote the growth of friendly bacteria in your gut while curtailing colonies of harmful bacteria. Technically, prebiotics are known as "fructooligosaccharides." Yes, that's a mouthful, so they've been nicknamed FOS for short. This carbohydrate is found naturally in many foods, including bananas, tomatoes, Jerusalem artichokes, onions, garlic, and various whole grains. If you eat a variety of these foods, you take in approximately 800 milligrams of FOS a day. So beneficial is this carbohydrate that it is increasingly being added to certain foods, such as juices, to provide a health benefit beyond the traditional nutrients a particular food contains.

After reaching your colon, FOS becomes a "meal" for health-promoting bacteria (known as probiotics), believed to enhance healthy flora in the intestines, improve digestion, and prevent disease. In other words, these good bacteria feed on FOS, and this results in increasing their numbers in the colon. (Bad bacteria, incidentally, do not feed on FOS.)

Increasing your consumption of prebiotics, whether through food or supplements, offers several benefits. Prebiotics:

- Increase good bacteria while reducing bad bacteria.

- Stimulate the absorption of water and minerals.

- Improve digestive health.
- Prevent certain types of diarrhea.
- Protect liver function.
- Reduce cholesterol and high blood pressure.

Supplementing with Prebiotics

Many health experts believe that Americans do not consume enough FOS-containing foods and recommend that we need about three times more than we currently eat. That being so, these important prebiotics are added to certain foods and found in a growing number of dietary supplements.

Prebiotics are available supplementally in capsules and in loose powder form. They are also cropping up in dietary beverages such as Ensure and other liquid carbohydrate supplements. The recommended dosage for pure FOS is 2,000 to 3,000 milligrams a day. Prebiotics are generally safe; the only reported side effect from supplementation is gas and bloating.

Please Pass the Fiber: Prescription for Prevention

High-fiber diets maintain the health of your digestive system and help it run smoothly. For a diet to earn the title of "high fiber," it must supply between 25 to 35 grams of fiber daily. If you weigh 175 pounds or more, you would want to shoot for even more fiber, say 40 to 45 grams a day. From information in chapter 2, you know there are two types of fiber—soluble and insoluble—and you should try to include both types in your diet by eating a variety of fruits, vegetables, cereals, and whole grains.

Getting 25 to 35 grams, or more, of fiber in your daily diet is not difficult. Start your day with a high-fiber breakfast, and you will be well on your way to meeting your daily fiber objective without really thinking about it or doing bothersome calculations. Try to eat a cereal for breakfast that has at least 4 grams of fiber per serving. Add to that cereal a high-fiber fruit such as raspberries (8 grams per cup) or blueberries (also 8 grams per cup), and you can meet nearly half the recommended fiber intake first thing in the morning. The rest of the day, choose one or two more fruits, several serv-

HIGH-FIBER JUMP-STARTS

	High-Fiber Breakfast Cereals	Fiber Grams Per Serving
General Mills	Fiber One—1/2 cup	14
	Multi-Bran Chex—1 cup	6.4
	Total Raisin Bran—1 cup	5
	Oatmeal Crisp with	
	Almonds—1 cup	4
Kellogg's	All Bran with Extra Fiber—1/2 cup	13
	Fiberwise—1/2 cup	10
	All Bran—1/2 cup	9.7
	Bran Chex—1 cup	8
	Whole-Grain Shredded	
	Wheat—1/4 cup	8
	Cracklin' Oat Bran—3/4 cup	6.5
	Bran Flakes—3/4 cup	4.6
Post	100% Bran—1/3 cup	8
	Raisin Bran—1 cup	8
	Shredded Wheat and	
	Bran—1 1/4 cup	8
	Bran Flakes—3/4 cup	5
	Fruit & Fibre—3/4 cup	5
	Grape Nuts—1/2 cup	5
	Shredded Wheat Original—2	
	biscuits	5

ings of vegetables, a slice or two of high-fiber bread (it should have 3 grams of fiber per slice), and a serving of legumes, and you will automatically serve up a high-fiber day for yourself. The table above identifies several high-fiber cereals that are good choices for breakfast to get your day started right.

Heart-Healthy Carbs

The old news is that a low-fat diet helps reduce your risk of heart disease. The new news is that a low-sugar diet does, too.

It's true that the mainline dietary strategy for fighting heart trouble centers on cutting out fat-ridden foods. That's essential. But reducing fat alone does not necessarily decrease your risk of heart disease. You have to limit the amount of added, refined sugar you eat, too, and that's what we will be focusing on in this chapter.

For background, heart disease is the nation's leading killer. Nearly 2,600 Americans die of it each day, an average of 1 death every 33 seconds. And if you have diabetes, this can double or quadruple your risk; heart disease is the number-one cause of death among diabetics. Although heart disease might build up over many years, its impact can be quite sudden, often resulting in a heart attack.

The blood vessels that provide oxygen and nutrients to the heart muscle encircle the heart like a crown, which is why they're called "coronary arteries." Problems begin when fatty deposits called plaque build up or a clot lodges in the artery, narrowing the passageway and choking off blood flow. A symptom of plaque buildup is angina, or chest pain. It's usually one of

CLASSIC SYMPTOMS OF HEART DISEASE

- A squeezing sensation in the chest that feels like a fist clenching your heart
- A heaviness in the chest
- Pain that radiates from the chest into the neck, shoulders, stomach, or back and doesn't go away
- Shortness of breath
- Intense sweating
- Fainting

the first warning signs of coronary heart disease. (Other symptoms are listed in the following box.) A heart attack occurs when blood flow to the heart muscle is cut off for a long period of time and part of the heart muscle dies.

With heart disease, prevention is paramount, and one of the many dietary measures you can take is to restrict the amount of sugar in your diet, and replace it with more healthy, wholesome carbohydrates. (Some of the other major preventive steps are listed on the following page.) When I talk about "sugar," I'm referring to added sugar, the stuff that is used to sweeten soft drinks and incorporated into other foods. Added sugar is a bad carb that should be limited in your diet. As I mentioned earlier, the average person eats a half-pound or more of added sugar a day. That's too much if you want to keep your ticker in top condition.

Sugar and Your Heart

The link between the consumption of refined sugar and heart disease was discovered in the 1940s and 1950s, when researchers found that the prevalence of heart disease was highest in countries where people ate the most sugar. Since then, much more has been learned about sugar and its effect on the heart. Sugar is harmful to your heart for at least six reasons.

LIVING A HEART-HEALTHY LIFESTYLE

Preventing and managing heart disease involves mostly lifestyle changes, including diet, that can drastically reduce your risk. One is to quit smoking. Smoking is public enemy number one when it comes to heart health. It strains the heart by constricting the arteries and making it beat faster. It elevates blood pressure and increases poisonous carbon monoxide in the blood.

• High blood pressure is a risk factor too. Even with a diagnosis of mild high blood pressure, you triple your risk of heart attack. Many times you don't even know you have high blood pressure until you have it checked.

• If you're overweight, try to trim down. Extra pounds are hard on your heart. The higher your weight climbs, the greater your risk. Being overweight also raises blood pressure and cholesterol, which are risk factors themselves. Even a weight loss of just ten pounds can improve your risk status. Maintaining weight is important, too. Cycling up and down in weight puts you at a greater risk. So does your body shape. If you're built like an apple, with extra pounds around your middle, you're at a greater risk.

• A low-fat diet will help you lose weight, too. Plus, it helps control cholesterol—another major risk factor in heart disease. The higher the level of LDL cholesterol in your blood, the greater your chances that plaque will build up in your arteries. Eating too much saturated fat, like animal fats, butter, and high-fat dairy products, can drive your cholesterol levels way up.

• Talk to your doctor about aspirin therapy. Aspirin reduces inflammation in the body. Plus, it is an "antiplatelet drug," which means it prevents platelets—tiny clotting substances in blood—from abnormally clumping together in a process called platelet aggregation. Platelet aggregation triggers the formation of dangerous blood clots called thrombi (or thrombus in the singular).

Reason #1: Sugar Promotes Inflammation

Sugar weakens your body's resistance to bacteria, viruses, and yeasts—all of which cause inflammation in both the heart and arteries. Inflammation plays a role in the build-up of artery-clogging plaque and is now recognized as a major risk factor for heart disease.

Inflammation is a bodily immune response that is triggered when your body is under attack from germs and other invaders. The immune system presses cells into service to destroy the invaders. Inside blood vessels, these cells can pile up, forming lesions that might eventually rupture and lead to a heart attack if preventive measures, mostly involving lifestyle changes, are not undertaken.

You can be tested for inflammation with a simple blood test that measures C-reactive protein (CRP), a protein that increases with the amount of inflammation in your coronary arteries. High levels of CRP are now believed to be the strongest and most significant predictors of heart disease, heart attack, and stroke.

One way to reduce inflammation in your body is to cut back or avoid high-sugar foods, as well as fatty meat, another inflammation-triggering food. A low intake of fruits and vegetables sets the stage for inflammation; so does not exercising. With some simple dietary and lifestyle changes, you can go a long way toward halting inflammation so it does not damage your heart and blood vessels.

Reason #2: Sugar Elevates Triglycerides

Synthesized in the liver, triglycerides are the form in which fat travels through your bloodstream. When there is too much of this fat en route through your system, the excess gradually finds its way to your inner arterial walls, where it is deposited, eventually obstructing your arteries. Two major dietary factors are known to elevate triglycerides: eating too much saturated fat (the type found mainly in animal foods) and eating too much added sugar. Several other factors besides sugar and saturated fat can elevate your triglyceride levels: drinking alcohol, smoking, taking estrogen, and following a diet high in other bad carbs.

When you have your cholesterol checked and analyzed, your triglycerides are checked and measured also. Any number lower than 150 is considered normal for triglycerides; higher numbers are considered a risk factor for heart disease. The following table lists healthy and unhealthy ranges for triglycerides.

If you have high triglycerides, you might also have elevated LDL cholesterol, because the two often go hand in hand. LDL cholesterol is a type of fat produced in your liver, usually in response to an excess of saturated fat

TRIGLYCERIDE READINGS

Normal triglycerides	Less than 150
Borderline high triglycerides	150 to 199
High triglycerides	200 to 499
Very high triglycerides	500 or greater

Source: American Heart Association

in your diet. Essentially, too much saturated fat disrupts your liver's ability to break down cholesterol, so it churns out LDL cholesterol, which can form lesions inside your arteries. LDL cholesterol is considered the "bad" cholesterol. You might also have too-low levels of HDL cholesterol (considered the good kind). When triglyceride levels are too high, HDL cholesterol tends to fall.

How exactly does sugar increase triglyceride levels? The more sugar you eat, the higher and faster it increases glucose in your blood. This causes your pancreas to pump out more insulin to lower your blood sugar. Insulin is a fat-promoting hormone. Secreting a lot of it prevents your body from burning fat, including triglycerides, and consequently, your liver converts the glucose into triglycerides and releases them into your bloodstream. Sugar isn't the only food to trigger this reaction. Eating a lot of carbohydrates that are high on the glycemic index—in other words, foods that quickly raise blood sugar—also causes increased production of triglycerides.

As I explained in chapter 2, one of the most damaging types of added sugar is fructose, mainly in the form of high-fructose corn syrup, a refined version of fructose made from corn and found in many processed foods and so-called health foods. Fructose is fat-forming in the body because it is metabolized differently than other sugars are.

Reducing your triglycerides requires revamping your lifestyle to reduce the amount of sugar you eat (including high-fructose corn syrup), cut back on the saturated fat in your diet, avoid or limit alcohol consumption, and become more active. For your fat source, consider substituting monounsaturated fats such as olive oil or canola oil. In research, both have been shown to lower triglyceride levels.

Reason #3: Sugar Affects Concentrations of Protective HDL Cholesterol

A heart-friendly cholesterol is high-density lipoprotein, or HDL cholesterol. Its job is to pick up the bad cholesterol from the cells in the artery walls and transport it back to the liver for reprocessing or excretion from the body as waste. HDL cholesterol has been nicknamed the "good cholesterol."

Eating "slow" carbohydrates—those low on the glycemic index—might positively affect concentrations of this good cholesterol—so says a British study of the dietary habits of 1,420 men and women. The researchers discovered that those who had diets rich in low-glycemic index foods, such as beans, had higher levels of HDL cholesterol.

By contrast, diets consisting of higher-glycemic index carbs drive HDL levels down. The reason for this has to do with insulin. Continually choosing sugar and quick-digesting carbs or large amounts of these foods—which are high on the glycemic index—raises blood sugar and insulin levels. High insulin contributes to low HDL cholesterol levels.

If your cholesterol profile is out of whack—that is, too-low HDL levels and too-high LDL levels—you might try eating low-glycemic index carbs for a few months to see if this makes any difference. Support this dietary change with other lifestyle measures, such as following a diet low in saturated fat and getting regular exercise.

Reason #4: Sugar Influences Obesity

Having a weight problem raises your blood sugar, your cholesterol, and your blood pressure, all of which are major risk factors for heart disease.

If you want to get your weight under control and avoid the health problems related to being overweight and obese, you must definitely restrict your sugar intake. When eaten in excess, sugar has a tendency to be converted into body fat. Sugar is rapidly released into your system, driving your blood sugar up too high and giving you a quick "rush," followed by a fast "crash." The first organ to react to this sugar overload is the pancreas, which responds by secreting more insulin into your bloodstream. Insulin activates fat cell enzymes, facilitating the movement of fat from the bloodstream into fat cells for storage. Additionally, insulin prevents glucagon (a hormone that opposes the action of insulin) from entering the bloodstream, and glucagon is responsible for unlocking fat stores. The cumulative result of these interactions is the ready conversion of simple sugars to body fat. If you need to lose weight, see chapter 9 for an effective way to trim down and still enjoy carbohydrates.

Reason #5: Sugar Robs Your Body of Important B Vitamins

Vital for energy, B vitamins are involved in nearly every reaction in the body, from the manufacture of new red blood cells to the metabolism of carbohydrates, fat, and protein. In excess, sugar leads to deficiencies in all the B vitamins, which are essential for healthy arteries.

Certain B vitamins—namely folate, vitamin B_6, and vitamin B_{12}—are needed to prevent the buildup of homocysteine in the blood. Homocysteine is a harmful substance that causes the cells lining arterial walls to deteriorate, leading to heart attack or stroke. A deficiency of vitamin B_6, in particular, leads to atherosclerosis.

Another important heart-helper is the B vitamin niacin. It is involved in the regulation of cholesterol and triglyceride levels.

With the exception of vitamin B_{12} (found mostly in protein foods), whole grains and vegetables are rich sources of B vitamins. Make sure your diet serves up plenty of these foods to protect yourself against deficiencies.

Reason #6: Sugar Promotes the Formation of AGEs

When your body is subjected to high levels of glucose—which can happen if you're eating a lot of sugar as a matter of habit—sugar rushing into the bloodstream as glucose links up (or glycates) with proteins in the body. The by-products of this reaction are the compounds called advanced glycation end products (AGEs), which can inflict damage in your body. AGEs alter tissue proteins, weakening bodily tissue (including blood vessels), reducing the elasticity of tissues, and interfering with normal cellular functions. All these situations can conspire to increase your risk of heart disease.

LDL cholesterol can become glycated, too—a process that prevents the normal shutdown of cholesterol synthesis in the liver. As a result, cholesterol levels soar, increasing the risk of atherosclerosis (the narrowing and thickening of arteries). Also, AGEs generate health-damaging free radicals.

The bottom line: When blood glucose is chronically elevated, there is a glut of glycated cells and substances in your system. Your body can't handle the excess, and the risk of health problems, including heart disease, is bound to increase. Slash the sugar to slash your risk.

Kind-Hearted Strategies

Clearly, a poor diet is a major cause of heart trouble, but you can do something about it. To improve your heart health and reduce your risk of heart disease, keep an eye on the kind of carbohydrates you eat, along with watching your fat intake. For best protection:

- Reduce the amount of sugar you eat by avoiding bad carbs such as soft drinks, candy, pastry, and high-sugar cereals. Limit desserts to twice a week and try to stick to lower-fat desserts. Have fresh fruit for dessert more often. Drink unsweetened beverages. If you have a sweet tooth, use artificial sweeteners instead of caving in to your sugar urge. There's a new one on the market called Tagatose, and it is discussed at the end of this chapter.

- Limit your intake of fat-free commercial foods. These might be low in fat, but they are often loaded with added sugar and calories.

- If your triglycerides are high or your HDL cholesterol is too low, try eating carbs that are low on the glycemic index (rated 55 or less): barley, high-fiber cereals, whole-grain pasta, oat bran, oatmeal, apples, pears, oranges, beans, peas, and sweet potatoes. These carbs are absorbed more slowly into the bloodstream. Slower absorption restrains surges in insulin release and slows the production of triglycerides in the liver. Stay on a diet like this for a few months; then have your blood fats (triglycerides and cholesterol) tested to see if this dietary strategy works for you.

- Follow a high-fiber diet. Studies show that people who eat the most dietary fiber (in carbs such as high-fiber cereals, whole grains, fruits, and vegetables) have the lowest death rates due to heart disease. Although a high-fiber diet means between 25 and 35 grams of fiber a day, you can protect your heart with as few as 16 grams a day. That's the finding of one major study in which people who ate 16 grams of fiber daily had one-third the risk of dying of heart disease as those who ate less than 16 grams. The reason fiber works so powerfully is threefold. Soluble fiber (in foods such as oat bran, beans, and apples) is very effective at lowering cholesterol. Plus, fiber is believed to interfere with the formation of blood clots that can trigger heart attack and stroke. This type of fiber also keeps blood glucose in safe bounds. Just a half-cup to a full cup of beans daily can significantly cut your cholesterol and control your blood sugar.

- Improve the quality of carbohydrates you eat by shifting your diet to more good carbs and fewer bad carbs. In the following table, you'll find a list of carbohydrates that are healthiest for your heart. Eat a variety of these foods through the week.

- Balance the amount of carbohydrates you eat with other nutrients. The American Heart Association recommends the following nutrient balance: 55 percent of total calories from carbohydrates, 30 percent or less from fat, and 15 percent from protein.

- Consider going meatless (vegetarian) several times a week. Many studies have proved that a low-fat vegetarian diet not only prevents heart disease, but also can reverse it. In one study, a vegetarian low-fat diet, in which total fat was reduced to 10 percent of calories, along with daily exercise and a stress management program, had dramatic effects on the heart health of patients. After a year, 18 of the 22 patients in the study showed significant clearing in their arteries, plus a 91 percent reduction in the frequency of chest pains. In effect, they had reversed their heart disease.

Most people find it easier to phase into a vegetarian diet by becoming a partial vegetarian. You can accomplish this with a gradual reduction in the amount of meat you eat. For example, plan several meatless days each week or a meatless meal each day. Another strategy is to become a semi-vegetarian (one who eats dairy foods, eggs, poultry, and fish but no red meat) or a pesco-vegetarian (one who eats dairy foods, eggs, and fish but no other animal flesh) before moving on to the stricter forms of vegetarianism.

To help you make the switch, here are some additional pointers:

- For starters, eliminate red meat but continue to eat poultry and fish occasionally.

- Make complex carbohydrates the centerpiece of your meals, rather than an animal protein.

- Early on, incorporate one vegetarian meal a day into your menus, eventually phasing out flesh foods and making all meals vegetarian.

HEART-PROTECTIVE GOOD CARBS

Good Carbs	How They Work
Barley	Contains soluble fiber, which reduces the absorption of cholesterol, decreases its production in the liver, and increases its elimination in the stool.
Beans and legumes	Rich in soluble fiber, which reduces the absorption of cholesterol, decreases its production in the liver, and increases its elimination in the stool.
Broccoli	Contains vitamin C, which helps keep arteries elastic and blood from clotting abnormally; also high in glutathione, an antioxidant that reduces heart disease risk, plus lowers blood pressure.
Carrots, pumpkin, sweet potatoes, yams, and other orange vegetables	Rich in carotenoids, including beta-carotene. Carotenoids might halt damage to the lining of the arteries.
Citrus fruits and other vitamin C–rich foods	Rich in vitamin C, which helps maintain the integrity of both the blood vessels and the heart muscle.
Fruits and vegetables	May lower levels of C-reactive protein (CRP) in the body, a marker of how inflamed heart arteries might be. High levels of CRP are a strong predictor of heart disease and are associated with blood clots. Fruits and vegetables contain natural antioxidants that might stop the formation of AGEs.
Garlic	May prevent the oxidation of cholesterol—a free-radical-generated process in which fatty streaks form on the inside of artery walls, leading to narrowing of arteries.

(continued)

HEART-PROTECTIVE GOOD CARBS

Good Carbs	How They Work
Green leafy vegetables	High in folate (folic acid), which lowers levels of homocysteine, a substance in the blood that increases the risk of heart disease.
Magnesium-rich carbs (tofu, beans and legumes, whole grains, and green leafy vegetables)	Magnesium is vital for heart health. It helps prevent heart attack, angina, irregular heart beats, high blood pressure, low HDL cholesterol levels, strokes, and improves overall heart function.
Oats and oat bran	Contain soluble fiber, which reduces the absorption of cholesterol, decreases its production in the liver, and increases its elimination in the stool.
Onions	May prevent the oxidation of cholesterol—a free-radical-generated process in which fatty streaks form on the inside of artery walls, leading to narrowing of arteries.
Soy foods	Lower levels of harmful cholesterol. (Twenty-five grams of soy protein a day is recommended.) Tofu, or soybean curd, has been shown to lower cholesterol in more than thirty-seven studies.
Whole grains	High in vitamin B_6 (pyroxidine), which helps prevent atherosclerosis by lowering homocysteine in the body, preventing abnormal blood clotting, preserving elasticity of blood vessels, normalizing cholesterol levels, and lowering blood pressure.
Whole wheat (including bread)	Helps reduce levels of homocysteine and LDL cholesterol.

TAGATOSE

If you're cutting back on sugar yet hate the aftertaste and limitations of artificial sweeteners, you're in for a treat in the form of tagatose, a low-calorie natural sugar that has been recently approved by the FDA for use in foods, beverages, and other products.

Manufactured from lactose, a simple sugar found in milk, tagatose looks like sugar, tastes like sugar, and best of all, cooks like sugar. The major difference between tagatose and sugar is that tagatose has fewer calories (roughly 6 calories per teaspoon compared to 16 calories for sugar) and has a low glycemic response, meaning that it won't push blood sugar and insulin up to unhealthy levels. For that reason, tagatose will probably be widely used as a sweetener in foods formulated for people with diabetes.

In addition, this new sweetener has applications in weight control. In one study, diabetic and normal patients using the sweetener on a daily basis lost weight at a consistent but gradual rate. What's more, tagatose has been shown in research to be a prebiotic, promoting healthy bacteria in the intestines for better digestion. It does not cause cavities either, and might even reduce the risk of tooth decay.

Tagatose is expected to be added to a wide range of foods, including soft drinks, cereals, baked goods, ice cream, candy, and chewing gum. It will also be used in toothpastes, cough syrups, medicines, and cosmetics.

Studied for more than ten years, tagatose appears to have no side effects.

- Find vegetarian recipes you enjoy.

- Consider "substitute meats"—products made from soybeans that are formed to look and taste like hot dogs, sausage, ground beef, and bacon.

- For variety and good health, include many different foods in your diet, because no single food contains all the nutrients you need.

You should now have a better idea that heart disease is a very real threat but that you can do something about it before it's too late. Cut back on your intake of sugar and saturated fat, eat the right types of carbs, exercise, make other changes in your lifestyle, get regular checkups, and recognize the symptoms of heart disease. By taking these positive actions, you can live better, and quite possibly, a lot longer.

Carbs and Weight Control:
The Good Carb Diet

To eat carbs or not to eat carbs.

That is the question on everyone's mind—everyone who wants to lose weight and stay trim, that is. Nowhere have carbohydrates been more bashed as they have been in the arena of weight control. When it comes to overweight and obesity, carbs are the nutritional equivalent of *persona non grata*. Do they deserve such a bad rap? Read on; the answer to that question might surprise you.

Carbohydrates and Obesity

More than two-thirds of the U.S. population is now considered overweight or obese (weighing 20 percent or more than what is considered a healthy weight), and obesity has become a global epidemic. One of the major nutritional reasons for such widespread (excuse the pun) obesity in America is hiding on the carbohydrate side of the dietary equation. Since low-fat dieting hit the diet scene in the early 1980s, everybody started cutting the fat from their diets while increasing calories from carbohydrates. The diet mantra at the time was: Slash fat, eat carbs. We bought into the low-fat, high-carbohydrate message as hapless participants, but little did we know that we were setting ourselves up for big problems down the road.

Over time, some alarming blips appeared on the nutritional radar screen: People started getting fatter, not thinner. After the government's Food Guide Pyramid (a meal planning guide) was introduced in 1991, sermonizing the value of eating six to eleven servings of carbohydrates a day (including some bad carbs) while eating fats "sparingly," the number of overweight, out-of-shape Americans jumped a whopping 61 percent. We cut the fat from our diets but got wider bottoms and bigger bellies in the process.

Whoa, hold it right there—aren't you supposed to slash fat to lose weight? Isn't low-fat dieting the ticket to a slender life? Isn't it? Not necessarily, say nutritional scientists who have studied the relationship between the simultaneous rise in obesity and the rise in carbohydrate intake.

What appears to be largely responsible for the escalating obesity in our country is that we are eating an excess of bad carbs—foods such as white bread, baked goods, foods with added sugar, and other processed junk foods. The carbohydrate-to-fat ratio in the typical American diet has gotten way out of whack. Bottom line: Eating too much of the wrong kinds of carbs—the bad carbs—can make you fat.

The Pros and Cons of Low-Carbohydrate Dieting

Another prevailing reason for carbs' bad rap is that dieters everywhere have gone high-protein, low-carb bonkers. I'm willing to wager that there are millions of people right now who are following a high-protein/low-carb diet—and you might be one of the converts. Perhaps even your doctor recommended it. If so, that's bad advice and bad medicine, especially if you care about your health.

I realize that these diets are popular because they let you load up on steak, cheese, and other high-fat fare. But they prohibit many other health-protective foods; most of them are the good carbs I have been encouraging in this book. You know the spiel of these diets: few fruits and vegetables, few carbohydrates.

What's wrong with this type of diet? To begin with, it is very high in fat. Roughly 50 percent or more of its calories come from fat—a lot of it saturated (that's one of the worst of the bad fats). If you are worried about cancer (particularly colon cancer or prostate cancer) or heart disease, you would want to avoid eating this way, because high-fat diets have been linked to numerous cancers and cardiovascular problems.

Also, a diet like this might not supply enough of the calcium you need for good bone health. Calcium is practically flushed out of your body with a high-protein diet.

What's more, the fiber on this type of diet would be miniscule. Eating a low-fiber diet is harmful to your digestive health and is a real invitation to diverticulosis, kidney stones, and gallstones. You'll be popping lots of laxatives, for sure, while on such a diet. Other health-preventive nutrients, such as antioxidants and phytochemicals that are found mostly in fruits and vegetables, would be missing from this diet as well.

There are other penalties, too. On a low-carb diet, your body is deprived of its favorite fuel. You're low on gas and feel it, physically and mentally. In this de-energized state, accompanied by a gloomy mood, it becomes tough to stick to your eating program, and another attempt to lose weight could bite the dust.

Why Low-Carb Dieting Works

When you think about the number of nutrients that aren't being delivered and the health risks involved, you have to ask yourself why in the world would anyone follow such a diet?

The reason you and many other people diet like this is because *low-carbohydrate dieting works*. When you reduce your intake of carbohydrates, your body starts drawing on its fat reserves for energy. Although your body prefers to fuel itself with carbohydrates, it will use fat as a backup if need be.

If you consider that a good deal of obesity is a result of eating too many calories from bad carbs, cutting carbs to lose weight just makes sense. But there's more to it: When you overindulge on bad carbs, your body overproduces insulin in an effort to transport glucose into cells. Medically known as "hyperinsulinemia," high concentrations of insulin trigger your body to create more fat cells. Another problem brought on by a diet high in bad carbs is "insulin resistance," in which insulin can't do its job of processing sugar and fats for energy. Consequently, your body starts storing more fat than is normal.

So by eating fewer carbs, you keep your blood sugar levels low, and insulin functions as it should—no hyperinsulinemia, no insulin resistance. Metabolically, your body is able to tap into its fat stores for energy, your weight starts decreasing, and before long, you're lean as a hairpin. So you see: Low-carb dieting is a very efficient way to shed pounds.

Still, many people are asking this important question: *Isn't there some way I can lose weight fast and effectively and still eat some carbs?* Or asked another way: *Can't I have my carbs and eat them, too?*

Answer: yes!

You can have your carbs and eat them, too, if the carbs you select are good carbs—and you eat them in the right proportions. You see, where low-carb dieting goes astray is that it advocates cutting too many carbs—even good carbs! The real key to success with low-carb dieting is cutting out bad carbs and reducing fast carbs (high-glycemic carbs), but replacing them with good carbs you love and that won't count against you when it comes to shedding fat.

The Good-Carb Diet

What I'm advocating is a low-bad-carb diet, rather than an out-and-out low-carb diet. This approach is called the Good-Carb Diet, and it combines the principles of low-carb dieting with the principles of good-carb nutrition. When you follow this approach to eating, don't be surprised if you get your cholesterol and your triglycerides under control along with your weight, because shunning bad carbs helps normalize blood fat. Plus, with this method of weight control, you can lose weight effectively, without sacrificing vital nutrients, and energize yourself with a variety of wholesome, healthy carbs.

The Good-Carb Diet takes a five-part fat attack strategy.

1. Eat 120 to 150 Grams of Good Carbs a Day, Including Glycemically Acceptable Carbohydrates

Generally, your body requires roughly 200 grams of carbohydrates a day for peak metabolic functioning. But you can get by with a slight reduction in carbs and still lose weight effectively. The amount of carbohydrates you can eat each day—and encourage your body's fat-burning processes—is approximately 120 to 150 grams a day, or about 480 to 600 calories of good carbohydrates daily. In other words, you do not have to slash your carbohydrate intake to ridiculously low levels like 20 to 50 grams a day in order to drop pounds. You can do it with a healthy, nonsacrificial intake of 120 to 150 grams. That should be welcome news to you if you've been suffering under a very low-carb intake of 50 grams or less!

Among your best choices for weight control are natural, unprocessed carbohydrates ranked 55 or below on the glycemic index of foods. These foods take longer to break down and, consequently, your blood glucose and insulin levels stay relatively constant during the digestion process. This helps control your appetite and boosts fat-burning. High-GI foods, on the other hand, do the opposite. They trigger an insulin spike, causing your body to create more fat cells. High-GI foods also stimulate your appetite. You feel hungry so you're apt to eat more—and more of the wrong foods stimulating your appetite.

Choosing low-GI carbohydrates makes it possible to lose weight more easily. If you base your carbohydrate selections *mostly* on this nutritional element, you will lose weight more quickly than with the same caloric total of other carbohydrates. Refer to chapter 2 for a list of low-GI carbs, and consult the food lists in the following "Good Carbs and Weight Loss" section. Not every carb you eat on the Good-Carb Diet has to necessarily be a low-GI one; simply base a majority of your choices on these foods. Other carbs to choose are what I call "low-starch carbs": salad vegetables, broccoli, cauliflower, green beans, spinach, and so forth. These foods are low in calories and starch and are loaded with health-bestowing nutrients.

2. Increase Your Daily Intake of Fiber

There is an incredibly easy, no-willpower way to manage your weight with good carbs, one that most of us should be doing but aren't: Eat more fiber. More fiber in your diet will help transform your dieting efforts into something so simple and automatic. You'll be able to keep your weight under control without even working at it or making yourself crazy.

For review, fiber (also called "roughage") is the digestion-resistant portion of plant foods. Over the past thirty years, "epidemiologic" studies, in which investigators study large populations of people to see who gets disease and who doesn't, have found that diets low in fiber are linked to a higher risk of obesity. Dietary fiber has several main talents when it comes to weight control:

• **Fiber fills you up but not out.** Soluble fiber, in particular, slows the passage of food from the stomach to the small intestine. This tends to make you feel full after eating. One of the best high-fiber fill-up foods you can eat is oatmeal. Bran and other high-fiber cereals aren't bad, either. In a study conducted at the Veteran's Administration Center in Minneapolis, people who feasted on high-fiber cereal for breakfast ate 150 to 200 less calories at lunch compared to those who had low-fiber breakfasts.

Another high-fiber heavyweight is the soluble fiber pectin, plentiful in apples, oranges, and bananas. Pectin inflates after soaking up water in the stomach, making you feel full.

- **Fiber promotes satiety.** It takes longer to crunch down on and chew up fibrous foods, so your meals last longer. That's a plus, because it takes about twenty minutes after starting a meal for your body to send signals that it's full. With enough fiber at mealtime, you're less likely to stuff yourself and eat too many calories. What's more, fibrous foods add very few calories to your meals.

- **Fiber stimulates the release of an appetite-suppressing hormone.** The hormone in question is called cholecystokinin (CCK). Secreted by the upper intestines after you eat a meal, CCK acts on nerves in your stomach and slows the rate of digestion. CCK is also released by the hypothalamus, the body's appetite control center. The net effect of CCK's release in your body is to tell your brain you're full. As a result, you feel full during your meal, so again you're less tempted to overeat.

- **Fiber controls your fat and sugar intake.** Fiber-rich diets tend to be low in fat-forming foods such as fats and sugar. By filling up on wholesome, fibrous foods, you have less room for foods that contribute to fat gain.

- **Fiber has a fat-binding effect.** Although fiber slows down the digestion of protein and carbohydrates, it does not do the same with fat. In the digestive system, fibers naturally bind to fats you eat and help escort them from the body, leaving fewer calories left to be stored as body fat.

- **Fiber helps regulate blood sugar.** High-fiber foods require prolonged breakdown and, thus, release blood sugar more slowly. This action helps prevent dips in blood sugar—dips that can lead to food cravings. A high-fiber diet helps maintain even energy levels throughout the day.

- **Fiber produces an automatic caloric deficit.** If you take in 35 grams of fiber a day (recommended for weight loss), you can count on expending nearly 250 calories from the total calories you ate that day. Because fiber makes your body work hard to process it, this actually creates a caloric deficit. You're burning extra calories, but without any extra effort.

- **Fiber increases transit time through your digestive system.** Transit time refers to how fast food moves from entry to exit over the course of digestion and absorption. Fiber in your diet speeds things along, meaning fewer calories are left to be stored as fat.

Fortunately, the carbohydrates you'll be selecting are loaded with this wonder nutrient fiber. The recommended dietary intake for fiber is between 25 and 35 grams a day. Bumping up your intake to the higher end of that range (35 grams) is recommended for weight loss. But do this gradually. If you start eating 35 grams of fiber tomorrow, you will feel bloat, gas, and other intestinal discomfort. Also, drink eight to ten glasses (8 ounces) of plain water every day while following a high-fiber diet. Fiber requires water for proper digestion and elimination.

If you have trouble increasing your fiber, adding a fiber supplement such as Metamucil is an easy way to obtain an additional 6 grams of fiber a day. However, fiber supplements should never replace high-fiber foods. It is always preferable to increase your fiber naturally through good-carb, fiber-rich foods.

3. Balance Your Ratio of Carbohydrates with Other Nutrients in Fat-Burning Proportions

The ratio of carbohydrates, proteins, and fat you eat on a daily basis is very important to effective weight control. A fair body of research indicates that you can lose weight most efficiently when your nutrients are proportioned in approximately the following manner:

- *Carbohydrates:* 40 percent of total daily calories

- *Protein:* 30 percent of total daily calories

- *Fat:* 30 percent or less of total daily calories

In a typical 1,200-calorie weight-loss plan, this method of losing fat breaks down in the following manner:

1,200-Calorie Daily Good-Carb Plan

Good carbohydrates: 480 calories from carbohydrates/120 grams of carbohydrates daily. (You can go as high as 150 grams, however, and still lose weight at a steady rate.)

Lean proteins: 360 calories from protein/90 grams of protein daily
Healthy fat: 360 calories from fat/40 (or less) grams of fat daily

4. Modify Your Calories

Although it seems out of "dieting vogue," cutting calories still works wonders for shedding pounds. To lose 1 pound a week (the maximum safe rate of weight loss), you'll need to eat 500 fewer calories a week. If you add calorie-burning exercise to this equation, your weight loss will be even greater. Most people can lose weight safely on a diet that provides 1,200 to 1,500 calories a day.

5. Restrict Starchy Carbohydrates After Your Midday Meal

There is an easy way to trick your body into thinking it is on a low-carb diet—and burn fat in the process: Cut back on certain types of carbs (those with higher-starch content) in the late afternoon and evening. This helps shift your body into a fat-burning mode.

Carbohydrates should be plentiful at breakfast, mid-morning snack time, and lunch meals to provide adequate fuel for the day's activities. But limit carbs at night, when they are less likely to be fully burned from exercise or activity.

Good Carbs and Weight Loss

Use the following foods to build your meals and lose weight.

The Foods

Low-Glycemic Starches (2 servings daily)

CEREALS AND GRAINS

All-Bran cereal	Long-grain rice
All-Bran with Extra Fiber	Oat bran
Bulgur wheat	Old-fashioned oatmeal

Parboiled rice

Pearled barley

Pumpernickel bread

Stone-ground
whole-wheat bread

Special K cereal

Whole-grain pasta

Whole-wheat crackers

BEANS AND LEGUMES

Black beans

Chickpeas

Kidney beans

Lentils

Lentil soup

Lima beans

Pinto beans

Soybeans

Split peas

VEGETABLES

Peas

Sweet potatoes

Yams

One serving = ½ cup cooked cereal, rice, or pasta (with ready-to-eat cereals such as Special K, a serving often counts as 1 cup, so read the labels to determine the appropriate serving); 1 slice bread, 1 medium sweet potato or ½ cup mashed sweet potato; ½ cup legumes or peas; or 4 high-fiber, low-fat crackers.

Low-Starch Vegetables (3 to 5 servings daily)

Alfalfa sprouts

All salad vegetables

Arugula

Asparagus

Bamboo shoots

Bok choy

Broccoli

Brussels sprouts

Cabbage

Cauliflower

Celery

Collard greens

Cucumber

Eggplant

Endive	Scallions
Green beans	Spinach
Jalapeño or other hot pepper	Summer squash
Leeks	Swiss chard
Lettuce, all varieties	Tomato
Mushrooms	Tomato juice
Parsley	Turnip greens
Pepper, sweet	Water chestnuts
Radishes	Wax beans

One serving=1 cup raw vegetables, ½ cup cooked vegetables, or 1 cup vegetable juice.

Glycemically Acceptable Fruits (2 to 3 servings daily)

Apples	Oranges
Bananas	Peaches
Blackberries	Pears
Blueberries	Plums
Cherries	Prunes
Dried apricots	Raspberries
Grapefruit	Strawberries
Grapes	Tangerines
Kiwifruit	Tangelos
Melons	

One serving=1 medium apple, banana, or orange; ½ cup berries; ½ grapefruit; ½ cup frozen or canned (in juice or water) unsweetened fruit; or ¼ cup dried fruit.

Lean Protein (2 servings)

Chicken (skin removed)	Reduced-fat lunch meats
Eggs or egg whites	Shellfish
Fish	Tofu
Lean cuts of beef	Turkey (skin removed)

One serving=3 to 4 ounces chicken, turkey, fish, shellfish, or lean beef; 2 slices reduced-fat lunch meat; 1 egg; 4 egg whites; or 2 ounces tofu.

Light Dairy Foods (2 servings)

Cottage or ricotta cheese, low-fat	Skim milk
Low-fat milk	Soy milk
Low-fat yogurt	Reduced-fat cheeses
Nonfat yogurt	

One serving=1 cup skim milk, low-fat milk, or soy milk; 1 cup low-fat or nonfat yogurt; 1/4 cup cottage or ricotta cheeses; or 2 ounces reduced-fat cheeses.

Healthy Fats (1 serving)

Butter	Reduced-fat or low-calorie salad dressing
Margarine (choose "trans-free" products)	
	Salad dressing
Mayonnaise	Unoiled nuts and seeds
Reduced-fat margarine or mayonnaise	Vegetable oil (olive or canola oils are best)

One fat serving=1 tablespoon margarine, butter, mayonnaise, or vegetable oil; 1 tablespoon salad dressing; 2 tablespoons reduced-fat or low-calorie salad dressing, margarine, or mayonnaise; 1 tablespoon unoiled, unsalted nuts and seeds. (You may use nonfat salad dressings more liberally— 3 to 4 tablespoons per day, if you wish.)

Using the above food lists, here is a "template" to help you plan your daily menu:

Breakfast

1 serving of a high-fiber cereal

1 serving light dairy

1 serving of a lean protein (optional for breakfast)

1 serving fruit

Snack

1 serving light dairy

1 serving fruit

Lunch

1 serving lean protein

1 serving low-starch carbohydrate

1 serving low-glycemic carbohydrate

1 serving healthy fat

Dinner

1 serving lean protein

1–2 servings low-starch carbohydrate

Sample Seven-Day Plan

Monday

BREAKFAST

½ cup cooked oatmeal

4 scrambled egg whites

1 cup skim milk

½ grapefruit

SNACK

1 cup nonfat yogurt mixed with 1 cup blueberries and 1 tablespoon low-calorie fruit preserves

LUNCH

Tuna salad made with 4 ounces water-packed tuna, 1 cup lettuce, 1 tomato slice, and 2 tablespoons low-calorie Italian salad dressing

1 cup nonfat vegetable soup

½ cup long-grained rice

DINNER

4 ounces roasted chicken breast, skin removed

1 cup asparagus

1 cup spinach

Nutritional information: 1,100 calories; 120 grams carbohydrate; 112 grams protein; 23 grams fat; and 20 grams fiber.

Tuesday

BREAKFAST

½ cup cooked oat bran

1 cup skim milk

1 banana

SNACK

1 cup nonfat yogurt mixed with 1 cup strawberries

LUNCH

Spinach salad made with 2 ounces reduced-fat feta cheese, 1 cup chopped raw spinach, assorted raw chopped salad vegetables, ½ cup garbanzo beans, and 2 tablespoons low-calorie French dressing

DINNER

4 ounces roasted turkey breast, skin removed

1 cup broccoli, cooked

1 cup summer squash, cooked

Nutritional information: 1,100 calories; 145 grams carbohydrate; 78 grams protein; 35 grams fat; and 31 grams fiber.

Wednesday

BREAKFAST

1 slice high-fiber toast

1 egg, poached

1 cup low-fat milk

1 orange, medium

SNACK

1 cup nonfat yogurt

1 cup sliced fresh raspberries

LUNCH

4 ounces roasted chicken breast, skin removed

1 tossed salad

Salad dressing prepared with 1 tablespoon vegetable oil and 2 table-spoons vinegar

½ cup mashed sweet potatoes

1 kiwifruit

DINNER

4 ounces baked cod

1 cup cauliflower, cooked

1 cup yellow wax beans, cooked

Nutritional information: 1,200 calories; 121 grams carbohydrate; 90 grams protein; 33 grams fat; and 25 grams fiber.

Thursday

BREAKFAST

½ cup All-Bran with Extra Fiber

1 cup soy milk

1 orange, medium

SNACK

Smoothie, blended with 1 cup low-fat milk and 1 cup frozen blueberries

LUNCH

Open-faced sandwich, prepared with 2 slices reduced-fat turkey ham, sliced tomato, lettuce leaf, 1 tablespoon brown mustard, and 1 slice high-fiber bread

1 cup nonfat vegetable soup

½ cup sliced peaches, water-packed

DINNER

5 ounces grilled salmon

1 tossed salad with 2 tablespoons low-calorie French dressing

1 cup turnip greens, cooked

Nutritional information: 1,100 calories; 134 grams carbohydrate; 72 grams protein; 34 grams fat; and 39 grams fiber.

Friday

BREAKFAST

1 cup Special K

1 cup low-fat milk

1 cup fresh raspberries

SNACK

4 whole-wheat crackers, spread with ¼ cup low-fat cottage cheese

LUNCH

Chicken Caesar salad, prepared with 3 ounces chicken breast, 1 cup chopped romaine lettuce, 1 sliced small green pepper, 3 tablespoons chopped onion, and 1 tablespoon Caesar salad dressing

½ cup fruit cocktail, water-packed

DINNER

4 ounces grilled rib eye steak

1 cup mixed vegetables

1 cup stewed tomatoes

Nutritional information: 1,232 calories; 144 grams carbohydrate; 83 grams protein; 40 grams fat; and 30 grams fiber.

Saturday

BREAKFAST

One egg, scrambled

1 slice high-fiber toast

1 cup low-fat milk

½ grapefruit

SNACK

1 cup nonfat yogurt

1 apple, medium

LUNCH

Chili prepared with ½ cup kidney beans, 3 ounces lean ground beef, and 1 cup cooked tomatoes with green chilies

1 tossed salad with 1 tablespoon low-fat salad dressing

1 cup melon balls

DINNER

5 ounces steamed shrimp

½ cup coleslaw prepared with 1 tablespoon low-fat slaw dressing

1 cup broccoli

Nutritional information: 1,200 calories; 129 grams carbohydrate; 100 grams protein; 35 grams fat; and 23 grams fiber.

Sunday

BREAKFAST

½ cup All-Bran with Extra Fiber

1 cup nonfat yogurt

1 cup fresh raspberries

SNACK

Soy smoothie blended with 1 cup soy milk and ½ cup frozen peaches

LUNCH

Chicken chef salad prepared with 4 ounces baked chicken, 1 cup chopped lettuce, 1 cup chopped assorted salad vegetables, and 2 tablespoons low-calorie French dressing

1 slice high-fiber bread

DINNER

4 ounces baked turkey breast, no skin

1 cup cooked summer squash

1 cup cooked broccoli

Nutritional information: 1,100 calories; 128 grams carbohydrate; 92 grams protein; 29 grams fat; and 47 grams fiber.

Boost Fat-Burning with Exercise

You can follow this approach to eating—or any diet, for that matter—and lose weight, at least for a time. But you will be so much more successful, especially over the long haul, if you add regular exercise to the weight-loss equation. Here are some guidelines for accelerating your weight loss through exercise:

- Start out slowly, with an easy-to-do activity such as walking. Aim for three to five hours a week.

- After two to three weeks, begin to gradually increase the amount of effort you put in. You can do this by increasing the duration of your exercise session, say from 30 minutes to 45 minutes. The longer you work out, the more fat you'll burn. Or you can increase the number of times you work out each week. If you've been exercising three times a week, gradually work up to four or five times a week. You'll burn more calories—and more fat.

- Add a strength-training component, such as weight training, to your workout. Weight training builds body-shaping muscle and, because muscle is the body's most metabolically active tissue, increasing it helps burn more fat. Weight training also preserves lean muscle, which can wither away when you diet.

- Perform "combination exercise." This involves a single workout session in which you perform strength training, followed by some form of aerobic activity. This sequence—weight training first, aerobic activity second—shifts your body into a fat-burning mode. Lifting weights naturally forces your body to draw on stored muscle glycogen for fuel. During a 30- to 45-minute training session, you can use up a lot of glycogen. Afterward, your body is glycogen-needy—the perfect time to start your aerobics. Theoretically, your body then starts drawing on fatty acids for energy during the aerobics. You'll burn more fat, and get more trim as a result.

Enjoy the Results

After several weeks of eating and exercising according to this plan, you'll be slimmer, stronger, and healthier. And you will do it without sacrificing carbohydrates. When you reach your goal weight, gradually add carbohydrates back into your diet so you reach the 200-gram-a-day requirement. But always choose good carbs, using the scorecards in chapter 2. As for calories, research indicates that women can maintain their weight with a daily caloric intake of 1,400 to 1,900 calories a day; men, 2,000 to 2,500 calories or more, depending on activity levels. Your best bets for keeping your weight off include exercising on a regular basis and selecting natural, unprocessed high-fiber carbs in amounts that will keep unhealthy, excess pounds banished for good.

Carb Power for Exercisers and Athletes

If you are an exerciser or athlete, your goal is to maximize your stamina, energy, and performance in order to be the very best you can be in your particular sport or physical endeavor. What is the optimum way to achieve that goal? In a word, *carbohydrates*. The key to push-yourself-to-the-limit performance lies in your body's chief energy fuel, carbohydrates.

The power of carbohydrates to improve stamina and boost performance has been known for well over a century. In fact, the carbohydrate/athletics story began as far back as 1887, when two French scientists observed that the jaw muscle of a horse rapidly took up glucose when the animal began chewing its food. In other words, the glucose gunned the horse's muscle. The rest, as they say, is history. Since then, carbohydrates have been considered the most important source of energy for fueling the active body. Consider the following:

- When you work out intensely during exercise, training, or competition, carbohydrates supply 80 to 95 percent of the fuel you burn.

- Unless your body is well stocked with carbohydrates, it will break down its own muscle for energy, and you will loss muscle mass, strength, and power.

- If you do not consume enough carbohydrates during exercise, your performance will slow down by as much as 60 percent of your normal capacity.

- As the primary fuel for your brain and nervous system, adequate carbohydrates give you the nutritional wherewithal to make quick decisions and fast reactions—both necessary to excellent performance in competitive athletics.

Like a gas tank with its limited number of gallons, your body has a limited capacity to store carbohydrates. Your muscles, for example, can store about 1,200 calories of glycogen; your liver, about 400 calories. If there were no other fuel source available during exercise (like fat), carbohydrates would provide enough nutritional "gas" for a 32-kilometer run, or about 20 miles. Thus, you must be diligent about refilling your body with sufficient carbohydrates. If you do so, you can train harder on successive workouts for greater gains.

The amount of glycogen you have in your muscular tanks is directly related to how much carbohydrate you eat and how well trained you are. Highly trained athletes, for example, are very efficient at storing muscle glycogen. Naturally, the more glycogen you can store in your muscles, the longer you can train or work out. So to achieve peak physical performance if you are an athlete or a serious exerciser, your diet should be higher in good carbs (at least 65 to 70 percent of your total daily calories) than what is recommended for less-active people. That being the case, let's look at specific carbohydrate strategies you should follow to ensure winning performance.

Strategy #1: Stay Well Fueled with Carbohydrates

To keep your muscles stocked with glycogen, stay on a carb-plentiful diet from week to week, one that is designed mostly with the many good carbs we have discussed throughout this book. This is the best, all-around strategy to follow for consistently high performance. Case in point: A study of hockey players, whose sport requires both muscular strength and aerobic endurance, found that during a three-day period between games, a high-carb diet caused a 45 percent higher glycogen refill than a diet lower in carbs. By consistently fueling yourself with carbs, you can keep your muscles well stocked with glycogen.

What counts as a high-carb athletic diet? Exercise scientists recommend that if you are a hard-training exerciser or athlete, you should eat between 7 and 8 grams of carbohydrates per kilograms of body weight on a daily basis to keep your glycogen tanks full.

Let's say, for example, that you weigh 150 pounds. In kilograms, your weight would be 68 kilograms (150 divided by 2.2). Thus, your daily carbohydrate intake should be 476 to 544 grams (68 times 7 or 8). Again, that means about 65 to 70 percent of your total daily caloric intake should come from carbs.

Strategy #2: Practice "Carbohydrate-Loading" Prior to Competition

Used in competitive sports for more than thirty years, carbohydrate-loading is a technique of pushing more glycogen into your muscles than is normally available. Followed seven days prior to competition, carbohydrate-loading works best for endurance competitions lasting at least 60 to 90 minutes or more.

The scientifically accepted strategy for carbohydrate-loading is as follows:

- Seven days prior to your competitive event, purposely empty your glycogen stores with an intense training bout that lasts approximately 90 minutes.

- Over the next three days, consume a varied, mixed diet composed of 45 to 50 percent carbohydrates. Train at moderate intensities for 45 to 60 minutes. (For an example of how to plan a precompetition meal, see the following table.)

- During the last three days prior to the competition, increase your carbohydrate intake to 70 percent of your total daily calories. Train at moderate intensities for 30 to 45 minutes. (Refer to the precompetition meal plan in the following table.)

- Two to four hours prior to the race, consume a high-carbohydrate meal. Ideally, eat about 100 to 200 grams of carbs, or roughly 400 to 800 calories. Choose solid or liquid foods, depending on what your system can best handle. Some examples include a combination of fruit, bread, rice, or pasta, with skim milk or nonfat yogurt. Other excellent prerace carbs include liquid meal replacers such as Boost or Ensure, or a product

PRECOMPETITION DIET PLANS

	45 to 50% Carbohydrates	70% Carbohydrates
Breakfast	1 slice whole-wheat toast 3 egg whites, scrambled 1 cup fresh blueberries 1 cup low-fat milk	1 cup Cream of Wheat cereal 1 cup low-fat milk 2 medium bran muffins ½ grapefruit
Snack	1 high-protein bar	1 cup fruit-flavored yogurt
Lunch	4 ounces cooked reduced-fat ham Potato salad: 1 medium boiled potato, 1 small celery stalk (chopped), 2 tablespoons low-fat mayonnaise, and 1 teaspoon yellow mustard 1 medium apple	Tuna sandwich: 2 ounces water-packed tuna, 2 slices whole-wheat bread, and 2 tablespoons low-fat mayonnaise 2 medium carrots 1 banana
Snack	1 cup fruit-flavored yogurt	1 sports bar containing carbohydrates and protein
Dinner	5 ounces grilled salmon 1 cup steamed and chopped broccoli 2 cups tossed salad 2 tablespoons reduced-calorie dressing	2 cups whole-wheat pasta 1 cup spaghetti sauce with 4 ounces cooked extra-lean ground turkey 2 cups tossed salad 2 tablespoons reduced-calorie dressing 1 cup fruit cocktail, water-packed ½ cup frozen nonfat yogurt

like GatorLode, a high-carbohydrate-loading source. GatorLode and similar products are formulated with carbohydrates only (no protein) and some vitamins. Be sure to drink 16 ounces of fluid two hours prior to your race to make sure you start the competition in a well-hydrated state.

Strategy #3: Incorporate Carbohydrate Supplements into Your Diet

Your personal fitness and training program should allow for the use of carbohydrate supplements. These include sports drinks with added electrolytes; carbohydrate meal replacers formulated with protein; and sports bars containing carbohydrates, proteins, and other nutrients. Carbohydrate supplements like these are designed to replenish nutrients lost during exercise, provide energy, or enhance muscle growth. Further, the performance-enhancing advantages of these supplements are backed by impressive research. Some of the most popular and effective carbohydrate supplements are discussed in the following sections.

Sports Drinks (Fluid-Replacers)

Also known as fluid-replacers, these beverages contain mostly water (about 85 percent), but are formulated with electrolytes and small amounts of carbohydrates. Electrolytes are dissolved minerals; their job in your body is to conduct electrical charges that let them react with other minerals to relay nerve impulses, make muscles contract or relax, and help carry water through the small intestine and into the bloodstream. Electrolytes can be lost through sweat during hard workouts, athletic competitions lasting an hour or longer, or heavy on-the-job physical labor, and need to be replenished.

Sports drinks do a number of things: They supply energy for working muscles, enhance performance, stimulate rapid fluid absorption, encourage better fluid consumption due to their sweet taste, and replace water and electrolytes lost through sweat.

Most sports drinks are formulated with about 6 to 8 percent carbohydrates. The main forms of carbohydrates used in these supplements are glucose, fructose, and sucrose—a combination shown in research to be very effective at accelerating energy to your muscles.

Fructose, however, isn't as fast at furnishing muscle glycogen as glucose is. What's more, fructose can cause digestive problems if it is the only carbohydrate used in a sports drink. Even so, fructose is beneficial in promoting rapid resynthesis of liver glycogen. That's important, too, because whereas muscles tap into their own private stock of glycogen for energy, the liver stores glycogen for use by the entire body. This is one case in which fructose earns respect as a good carb.

Sports drinks are very safe when used as directed. It is best to avoid products that are carbonated or that contain caffeine. Carbonated beverages release their carbonation (carbon dioxide) in the stomach, which increases the risk of bloating, stomachache, and nausea. Caffeine acts as a diuretic, which can be detrimental during endurance competition.

Carbohydrate Meal-Replacers

Available as bars, in cans, or mix-it-yourself powders, meal-replacers are formulated to reproduce as closely as possible the nutrition you would get from food, complete with carbohydrates, protein, fat, vitamins, and minerals.

Meal-replacers serve as fortified snacks, as a supplement to a healthy diet, and as a performance-enhancing food used most effectively post-workout to aid in muscle recovery. Unlike sports drinks, they contain a higher amount of carbohydrates, usually about 60 grams or so. Meal-replacers are not to be confused with "carbohydrate-loaders," a type of liquid supplement formulated with only carbohydrates and no protein.

As to which is better for an endurance athlete, bar or beverage, it depends on personal choice. Beverages are great if you can't tolerate solids before or following exercise. Bars, on the other hand, have advantages in certain sports, such as long-distance cycling. Many cyclists like to wrap taffylike meal-replacer bars around their handlebars. The bars stick without crumbling, and the cyclists can simply peel them off and munch when needed.

Because these products vary in nutrient composition, try several brands to determine which works best for you. Never test-drive a new product at competition time; instead, experiment through trial and error during training.

Don't use meal-replacers in lieu of real meals to skimp on calories, at least not on a regular basis. They really aren't intended for that purpose, despite their name. Meal-replacers fail to provide all the nutritious food factors such as disease-preventing phytochemicals found in actual food. If you're dieting and want to use a meal-replacer as a snack, be sure to figure its calorie count in your daily allowance. Most of these products contain between 250 and 350 calories per serving.

Energy Gels

Energy gels are highly concentrated carbohydrates with a puddinglike consistency that are packaged in single-serve pouches. Designed for athletes and exercisers, these products provide 70 to 100 calories and 17 to 25 grams of carbohydrates per serving. The carbohydrate source is usually a mixture of simple carbs (usually fructose or dextrose) and maltodextrin or corn-starch. Some of the leading products on the market include ReLode, Clif Hot Shot, Pocket Rocket, and Squeezy.

Quickly absorbed into the bloodstream, energy gels are a good source of immediate food energy, particularly during extended exercise efforts. Experts generally advise that if you're an endurance athlete, you need to refuel yourself during exercise with 50 to 60 grams of carbohydrates each hour. That being so, take one energy gel packet every 30 minutes during a long race, bicycle ride, or other endurance event.

These supplements generally contain no fiber or other nutrients and, therefore, should not be used to replace the good carbs in your diet. When using energy gels, make sure you take in sufficient water to process the carbohydrates and prevent dehydration.

Wise Use of Carbohydrate Supplements

Always think of these products as supplements, or additions, to your diet, rather than substitutes for wholesome, natural carbs. Ultimately, the best way to fuel and nourish your body is by eating a varied, nutrient-rich diet of low-fat proteins and dairy products, fruits, grains, and vegetables.

Strategy #4: Time Your Use of Carbohydrates, Including Supplements

As critical as the amount and type of carbohydrates you consume is the timing of your intake. There are three critical periods in which to consider consuming carbohydrates: before your workout, during your workout, and after your workout.

Preworkout:

If you're an endurance athlete, use carbohydrate meal-replacers as pre-exercise snacks to boost calories for extra energy. You can also use meal-replacers for carbohydrate-loading, in addition to other carb sources, to keep your glycogen reserves well stocked for competition.

If you're a strength trainer in a mass-building phase and want to push to the max, fuel yourself with carbohydrates before your workout. Eat a low-fat,

SUPPLEMENTAL RIBOSE

What It Is: Supplemental ribose is a relatively new supplement available in tablets, nutritional drinks, energy bars, or in powder form. It may help boost your pep by revving up certain energy-producing systems in your body. Ribose occurs naturally in your body anyway, and so taking in some extra may give you an edge.

For background, your body's own ribose is found in every cell. It is a beneficial simple sugar that forms the backbone of DNA and RNA, the genetic material that controls cellular growth and reproduction.

Ribose stimulates your body's production of ATP (adenosine triphosphate), the main energy-releasing molecule of all living cells. Cells need ATP to function properly. Your heart uses ATP to beat, for example, and your muscles use it to move.

Normally, your body can produce and recycle all the ATP it needs, especially when there is an abundant supply of oxygen. But under certain circumstances—namely ischemia (lack of blood flow to tissues) and strenuous exercise—ATP cannot be regenerated fast enough, and energy-producing compounds called adenine nucleotides may be lost from cells. This can impair muscle function and tax strength, because cells need adenine nucleotides to produce sufficient amounts of ATP.

Proof that intense exercise can deplete cells of energy-producing nucleotides was demonstrated dramatically in a study published in the *Journal of Applied Physiology* in which eleven healthy men engaged in high-intensity training on a cycle ergometer three times a week for six weeks, then afterward trained twice a day for a week. Another group (nine men) rested for six weeks, then trained twice a day with the others. In the first group, nucleotide content of skeletal muscle dropped by 13 percent; in the second group, by 25 percent. In both groups, nucleotide levels had not fully recovered even after seventy-two hours of rest.

Ribose supplementation may restore these lost nucleotides. In one animal study, ribose supplementation increased the rate of nucleotide synthesis in resting and exercising muscles by three to four times. Other animal studies have found that ribose can restore energy levels to near normal within twelve to twenty-four hours. Studies are ongoing to confirm that ribose may be equally as beneficial in humans.

To preserve high cellular levels of ATP, the recommended dosage of supplemental ribose is 3 to 5 grams daily as maintenance dose, and 5 to 10 grams daily if you are a competitive athlete.

Ribose supplements appear to be safe, although at mega doses (60 grams a day), it causes gastrointestinal problems. If you have a heart condition or other serious medical problem, consult your physician before supplementing.

high-carb meal two to three hours prior to working out. In addition, make sure you are always well hydrated and drink 4 to 8 ounces of fluid immediately before exercise. Following this pattern will ensure you gain the greatest energy advantage from your pre-exercise meal without feeling full while you exercise.

Liquid meal replacers are a terrific pre-exercise carb choice. In a study of strength trainers, one group consumed a carbohydrate drink just before training. Another group was given a placebo. For exercise, both groups did leg extensions at about 80 percent of their strength capacity, performing repeated sets of 10 repetitions with rest between sets. What the researchers found was that the carbohydrate-fed group outlasted the placebo group, performing many more sets and repetitions.

During your workout:

If you want a little extra boost to improve your endurance during training or competition, supplement with a sports drink (like Gatorade) during your workout. For example, sip a half-cup of a sports drink every fifteen to twenty minutes to extend your endurance. This helps stabilize your blood sugar levels and maintain fluid levels and gives you extra pep. Research has consistently shown that drinking a sports drink during competition delays fatigue. By supplementing with a sports drink during competition, you spare your existing muscle and liver glycogen stores. That means you can run, bike, or swim longer because you're fueling yourself with a supplemental source of carbs so your body doesn't run out of steam.

Strength trainers can also benefit from a carb boost during training. Case in point: In one study, exercisers drank either a placebo or a 10 percent carbohydrate sports drink immediately prior to and between the fifth, tenth, and fifteenth sets of a strength training workout. They performed repeated sets of ten repetitions, with three minutes of rest between each set. When fueled by the carb drink, they could do more total repetitions than when they drank the placebo, which all goes to show that carbs clearly give you an energy edge when consumed during a workout. The harder you can work out, the more you can stimulate your muscles to grow.

But if you're trying to lose body fat, you might want to forgo supplemental carbs during your workouts, because although carbs boost your power and stamina, they might keep your body from dipping into its fat reserves for energy. Your entire workout could be powered solely on carb fuel and never significantly tap into fat stores for fuel. By working out in a moderately carb-needy state (no preworkout carbs), you can theoretically force your body to start using more fat for fuel. The downside, though, is that you could run low on energy.

The key here is to consider your goals—mass-building or fat-burning—and listen to your body for signs of fatigue. Adjust your carb intake accordingly, depending on your goals and energy levels.

Be sure to have a sports drink handy if you're exercising or working in hot weather, too. That's when fluid loss is greater than any other time of the year. You can lose more electrolytes, too, although the concentration of these minerals in sweat gets weaker the fitter you are. You also burn more glycogen working out in the heat—another good reason to quench your body with a sports drink.

Post-workout:

You've just finished a super-intense workout. If you could zoom down to the microscopic level of your muscles, you'd be astounded by the sight. There are tears in the tiny structures of your muscle fibers and leaks in muscle cells. Inflammation is setting in, and like cellular medics, white blood cells are on the scene to mend the damage and fix the leaks. Over the next twenty-four to forty-eight hours, muscle protein will break down, and additional muscle glycogen will be used up.

These are some of the chief metabolic events that occur in the aftermath of a hard workout. And although they might look like havoc, these events are actually a necessary part of "recovery"—the repair and growth of muscle tissue that take place after every workout.

During recovery, your body replenishes muscle glycogen and synthesizes new muscle protein. In the process, muscle fibers are made bigger and stronger to protect themselves against future trauma. Actual muscle growth occurs not while you're exercising, but in the recovery period following your workout.

There is much you can do to enhance this recovery process, and carbohydrates are a must-have, because your muscles are most receptive to producing new glycogen within the first few hours following your workout. This is the period in which blood flow to your muscles is much greater, a condition that makes muscle cells thirsty for glucose. Muscle cells are also more sensitive to the effects of insulin during this time, and insulin promotes glycogen restoration. Therefore, consume carbs *immediately* after you work out. Canned meal-replacers or sports bars are an easy way to immediately refuel.

Continue to take in 50 to 100 grams of carbohydrate every two hours following your workout until you resume your normal meal patterns. Strive for a total intake of 600 grams from food and supplements within twenty-four hours after your workout.

Another benefit of this post-training snack is that it helps trigger the elevation of key hormones (insulin and growth hormone) involved in muscle growth, especially in the period right after exercise. Case in point: At the University of Texas in Austin, nine experienced male strength trainers were given either water (which served as the control), a carbohydrate supplement, a protein supplement, or a carbohydrate/protein supplement.

The subjects took their designated supplement immediately after working out and again two hours later. Right after exercise and throughout the next eight hours, the researchers drew blood samples to determine the levels of various hormones in the blood, including insulin, testosterone (a male hormone also involved in muscle growth), and growth hormone.

The most significant finding was that the carbohydrate/protein supplement triggered the greatest elevations in insulin and growth hormone. Clearly, protein works hand in hand with post-exercise carbs to create a hormonal environment that's highly conducive to muscle growth.

For building muscle, consume a meal-replacer with your meals to boost calories. It takes an additional 2,500 calories a week to manufacture a pound of muscle. Increasing your calories by about 350 a day with a meal-replacer is an effective way to achieve that caloric surplus.

Strategy #5: Select Mostly Fast-Fueling Carbohydrates

The best types of carbs for refueling are those with a moderate- to high-glycemic index (GI) rating. These carbohydrates are digested and absorbed more rapidly, resulting in faster glycogen resynthesis. Such carbs include sports drinks, meal-replacers (beverages or bars), bagels, potatoes, brown rice, raisins, corn, sweet potatoes, oatmeal, and oranges.

A word of caution: There is a drawback of high-glycemic index foods. They might produce a fast, undesirable surge of blood sugar. When this happens, the pancreas responds by oversecreting insulin to remove sugar from the blood. Blood sugar then drops to a too-low level, and you can feel weak or dizzy.

Low-glycemic index foods, on the other hand, provide a more constant release of energy and are unlikely to lead to these reactions. By mixing and matching low- and high-glycemic foods in your diet, you can keep your blood sugar levels stable from meal to meal. The watchword here is *moderation:* Don't overdose on high-glycemic index foods or beverages.

Stay Carb-Charged

Whatever your sport or exercise, the single most important dietary factor affecting your performance is the amount of carbohydrates in your daily diet. Eating a high-carbohydrate diet, with mostly good carbs, will keep your muscles well fueled so you can challenge yourself physically and improve your performance during your regular workouts. Stick with good carbs as the mainstay of your diet, and you won't believe how great you'll look and how strong you'll feel.

Part III

Special Topics

•ELEVEN•

Your Smart-Carb Strategy

When you really think about it, deciding which carbs to eat boils down to some basic common sense: The more natural a carb is, the better it is for your body, because your body uses pure, natural carbs so much more efficiently than it uses processed junk carbs. Pure carbs are bursting with nutrients, each put to use in building and healing the body. Processed carbs, on the other hand, are nutritionally bankrupt and associated with various health problems.

So the key is to make carb choices that make sense for your particular situation, for example: Does diabetes run in your family, or are you struggling with it now? Do you want to do everything you can to deter cancer? Has your doctor told you to manage your digestive health with a higher-fiber diet? Do you have high triglycerides or cholesterol? Do you need to lose weight? Are you an athlete who wants to enhance performance naturally? Are you suffering from a health problem that might be helped with good carbs?

Once you've answered these questions, you can begin an overhaul of your nutrition program in order to eat more healthfully. Depending on your personal health and fitness goals, you will require a certain amount of carbohydrates each day—usually 40, 55, or 65 percent of your total daily calories. You can easily compute your allotment using the following formula:

Caloric Level	Total Carb Grams in a 40% Carb Diet	Total Carb Grams in a 55% Carb Diet	Total Carb Grams in a 65% Carb Diet
1,200	120 grams	165 grams	195 grams
1,500	150 grams	206 grams	244 grams
1,800	180 grams	248 grams	293 grams
2,000	200 grams	275 grams	325 grams
2,200	220 grams	303 grams	358 grams
2,500	250 grams	344 grams	406 grams
3,000	300 grams	413 grams	488 grams

Multiply your *total daily calories* by *the percentage of total daily calories from carbs* divided by *4*, the number of grams in each gram of carbohydrate, to determine your *daily carb requirements* in grams.

Let's suppose, for example, that you want to determine your carbohydrate needs for a 2,000-calorie diet in which 65 percent of your calories come from carbs. Here's how the calculation works out:

- Multiply your total daily calories by 65 percent (.65). For example: 2,000 calories×.65=1,300.

- Divide your daily carb calories by 4, the number of grams in each gram of carbohydrate: 1,300÷4=325 grams of total carbohydrate for the day.

To make it even easier, I've included the following table, which lists amounts of carbs in grams for various caloric levels and percentages. You'll also find a table listing the carb count for many of the carbs discussed in this book.

	Food	Serving Size	Carbohydrates (grams)
Beverages	Apple juice	1 cup	21
	Carrot juice	1 cup	22
	Concord grape juice	1 cup	32
	Cranberry juice cocktail	1 cup	14
	Grapefruit juice	1 cup	23
	Pineapple juice	1 cup	29
	Tomato juice	1 cup	9
Breads	Bagel	1 medium	38
	Bran muffin	1 medium	18
	Branola (high-fiber)	1 slice	18
	Pita pocket	1 whole	31
	Pumpernickel bread	1 slice	10
	Rye bread	1 slice	10
	Stone-Ground whole-wheat bread	1 slice	14
	Whole-wheat bread	1 slice	12
Cereals	100% Bran	⅓ cup	23
	All Bran	½ cup	23
	All-Bran with Extra Fiber	½ cup	23
	Bran Chex	1 cup	39
	Fiber One	½ cup	24
	Granola	¾ cup	36
	Raisin Bran	1 cup	47
	Shredded Wheat and Bran	1¼ cup	47
	Special K	1 cup	23
	Oat bran, cooked	½ cup	41
	Oatmeal	½ cup	14
Dairy Foods	Low-fat milk	1 cup	11
	Skim milk	1 cup	13
	Yogurt, sugar free	1 cup	14
Diabetic Supplements	Choice DM bar	1 bar	19
	Choice DM beverage	8 ounces	24
	Glucerna bar	1 bar	24
	Glucerna beverage	8 ounces	29

(continued)

	Food	Serving Size	Carbohydrates (grams)
Fruits	Apple	1 medium	21
	Apricots, dried	6 halves	12
	Avocado	1 medium	15
	Banana	1 medium	28
	Blueberries	1 cup	20
	Cantaloupe	1 medium wedge	6
	Cherries	1 cup	70
	Figs, dried	2 figs	24
	Grapefruit	½ fruit	9
	Grapes	1 cup	16
	Kiwifruit	1 medium	11
	Orange	1 medium	15
	Papaya	1 whole	30
	Peach	1 medium	11
	Pear	1 medium	25
	Pineapple	1 cup chunks	19
	Plum	1 medium	9
	Prunes, stewed	1 cup	69
	Raspberries	1 cup	51
	Strawberries	1 cup	11
Legumes, Cooked	Black beans	½ cup	17
	Garbanzo beans	½ cup	17
	Kidney beans	½ cup	20
	Lentils	½ cup	13
	Lima beans	½ cup	20
	Navy beans	½ cup	17
	Peas	½ cup	12
	Pinto beans	½ cup	15
	Soy milk	1 cup	4
	Soybeans	½ cup	8
	Tofu	2 ounces	1
Pasta	Macaroni	½ cup	18
	Whole-wheat spaghetti	½ cup	18

(continued)

	Food	**Serving Size**	**Carbohydrates (grams)**
Sports Supple- ments	Gatorade	8 ounces	14
	GatorLode (Hi-Carb)	11.6 ounces	71
	PowerBar	1 bar	28
Vegetables	Artichoke, boiled	1 cup	19
	Beets, boiled	1 cup	17
	Broccoli, boiled	1 cup	8
	Broccoli sprouts	1 cup	10
	Cabbage, boiled	1 cup	7
	Carrot, raw	1 medium	6
	Corn, canned	1 cup	19
	Onions, raw, chopped	¼ cup	3
	Potato, baked	1 medium	51
	Red pepper, raw, chopped	¼ cup	2
	Romaine lettuce, raw	1 cup	2
	Spinach, boiled	1 cup	7
	Sweet potato, baked	1 medium	28
	Tomato, raw	1 medium	6
	Winter squash, baked	1 cup	30
Whole Grains	Brown rice	½ cup	21
	Bulgur wheat	½ cup	24
	Couscous	½ cup	18
	Quinoa	½ cup	58
	Wheat bran	2 tablespoons	4

In the following sections are several health scenarios that sum up specific carbohydrate recommendations for certain situations and conditions. Use them as general guidelines and customize them where necessary. Some of this information has been covered in previous chapters, but is repeated here to give you a quick summary for review.

Good-Carb Guidelines for Brain Fitness

Carbohydrates are the leading nutrient fuel for your brain, supplying glucose for mental alertness, a bright mood, and a sharp memory. A brain-healthy diet should supply 55 to 65 percent of its total daily calories from good carbs. Most of these carbs should come from fruits and vegetables (at least five servings a day), because these foods are rich in antioxidants and phytochemicals, both of which protect brain cells from damage.

Good-Carb Guidelines for Preventing Cancer

Eating a healthy diet is one of the most significant moves you can make to prevent cancer. Generally, your diet should supply:

- 55 to 65 percent of total daily calories from good carbs
- 25 to 35 grams of fiber daily

In addition, the American Cancer Society recommends the following nutritional guidelines:

- Choose most of the foods you eat from plant sources. These include fiber-rich carbs such as fruits, vegetables, whole grains, and legumes. These foods are packed with vitamins, minerals, antioxidants, and phytochemicals—all known to reduce your risk of cancer. Eat at least five servings a day of fruits and vegetables; this amount has been shown in scientific research to reduce the risk of cancer, particularly lung cancer and colon cancer.

- Limit your intake of high-fat foods, especially those from animal sources. High-fat diets are tied to an increased risk of cancers of the colon, rectum, prostate, and endometrium.

- Stay physically active and maintain a healthy weight. Exercising thirty or more minutes a day helps control your weight (obesity is a risk factor for cancer), cuts your odds of colon cancer, and helps enhance your overall health.

• Limit or avoid drinking alcoholic beverages. Drinking alcohol increases your risk of cancer of the mouth, esophagus, pharynx, larynx, and liver in men and women and breast cancer in women.

Good-Carb Guidelines for Diabetes

Diabetes is a complex disease, requiring treatment that involves diet, exercise, lifestyle adjustments, and, for many people, injectable insulin or oral diabetes drugs. Generally, the recommended diet is one that is moderate in good carbs, low in saturated fat, and high in fiber. This translates into:

• 40 to 55 percent of total daily calories from good carbs

• 25 to 35 grams of fiber daily

If you have diabetes, you should eat less sugar. Foods high in sugar include desserts, sugar breakfast cereals, candy, table sugar, honey, syrup, and soft drinks. Generally, 10 percent or less of your total daily calories should come from added sugar.

The timing of your meals is also very important for preventing hypoglycemia (low blood sugar) and feeling energetic. Therefore, eat multiple meals and snacks, and try to spread those meals throughout the day. It's best to try to space your major meals four to five hours apart, and leave two hours between meals and snacks to allow for adequate digestion. Your body works best when nutrients are replenished every few hours.

In addition, be consistent with when you eat and how much you eat. Eat meals and snacks at approximately the same time every day, and eat the same portion sizes each day. This will help you achieve a balance among food, medication, and activity.

If you have type II diabetes, concentrate on cutting fat in your diet, particularly saturated fat and cholesterol, as well as losing weight, because a high-fat diet is a risk factor for heart disease, the major complication of diabetes.

The same dietary strategies, namely the selection of good carbs in carefully controlled amounts, that are used to manage diabetes can also be employed to prevent it.

Good-Carb Guidelines for Digestive Health

The number-one recommendation for excellent digestive health is to eat a high-fiber diet, with 25 to 35 grams, or more, of fiber from good carbs. Because of its ability to add bulk to the diet and softness to the stool, fiber moves through your digestive system like a one-nutrient janitorial crew, cleaning out toxins, cancer-causing agents, and cholesterol. This attribute earns fiber its reputation for being the most effective remedy for constipation, as well as for reducing the risk of many serious digestive diseases, including diverticulosis and colon cancer. The best high-fiber carbs are beans, legumes, whole grains, fiber-fortified cereals, and certain fruits.

Good-Carb Guidelines for Heart Disease

To bring your triglycerides and cholesterol down, your carbohydrate intake should be 55 to 60 percent of your total daily calories. (This is the carb intake recommended for heart health by the American Heart Association.)

It is preferable to choose good carbs—vegetables, fruits, whole grains, and high-fiber cereals—over simple sugars. Good carbs supply more fiber, vitamins, minerals, antioxidants, and phytochemicals than foods high in added sugars. Research has shown that sugar might be a risk factor for heart disease because it elevates triglycerides, which are harmful to the heart. Thus, avoid products that list more than 5 grams of sugar per serving on the label. If the specific amount of sugar is unlisted, shun products with sugar listed as one of the first four ingredients on the label. Sugar goes by various other names, too: sucrose, dextrose, maltose, lactose, maltodextrin, corn syrup, and high-fructose corn syrup, to name just a few.

Strive to eat five or more daily servings of fruits and vegetables, and six or more servings of grains—bread, cereal, or rice, for example.

Other recommendations include the following:

- Restrict your fat intake to 30 percent of your total calories. Less than 7 percent of your total calories should come from saturated fats and fats called trans fats, which are found in stick margarine and vegetable shortening.

- Limit your daily intake of dietary cholesterol to less than 200 milligrams.

- Exercise for thirty or more minutes most days of the week.

If your triglycerides and cholesterol don't change for the better with diet and exercise, consult your physician. You might need a prescription medication to normalize your blood fats.

Good-Carb Guidelines for Weight Control

Being overweight, defined as 20 percent or more above your ideal weight, puts you in harm's way of numerous life-threatening diseases, among them heart problems, stroke, cancer, and diabetes.

Without question, it can be challenging to lose weight, especially if you're far from your ideal weight. But it's not impossible, either.

One of the most important steps you can take is to choose mainly good carbs in the following amounts:

- 40 percent of your total daily calories from good carbs (This amount is low enough to encourage your body to burn stored fat.)

- 120 to 150 grams a day of good carbs

- 25 to 35 grams of fiber daily (Eating 35 grams of fiber a day is most effective for reducing body fat.)

In addition, eat fewer calories than your body uses each day. To lose 1 pound of body fat, you have to create a 3,500-calorie deficit, either by eating less, exercising more, or both. By cutting your total calorie intake by 500 calories each day, for example, you should be able to lose 1 pound a week (500 calories×7 days)—a safe rate of weight loss. If you add exercise to this equation and burn any extra calories, your fat loss will be even greater. An hour of exercise, for example, can burn up anywhere from 250 to 500 calories.

Your weight-loss diet should be as low in fat as possible, because reducing dietary fat is one of the best ways to shed pounds. By keeping your total fat intake to 30 percent or less of your total daily calories, you might be able to lose body fat with less restriction in total calories.

In addition, try to curb your intake of fat-forming foods such as sugar, processed foods, and alcohol. By limiting these foods, you'll automatically reduce the number of calories in your diet.

Good-Carb Guidelines for Athletic Performance

Ideally, 60 to 65 percent of the calories in your daily diet should come from carbohydrates, particularly good carbs, and this requirement goes for most physically active people. The best way to increase good carbohydrates is to add foods like whole grains, whole-grain cereals and breads, potatoes, yams, and legumes to your diet. Feel free to use higher-glycemic carbs, such as sports drinks, to refuel and restock your muscles with glycogen for energy. Other carbohydrate supplements, such as those discussed in chapter 10, are useful for enhancing performance, endurance, and muscle growth.

Good Carbs for Life

As you read through these guidelines, you probably detected a single thread running through each: Natural, unprocessed good carbs—grains, vegetables, and fruits—are healthy for every part of your body. Put another way, what is good for your brain is good for your heart—and for practically every other organ and body system. Good carbs simply mean good health.

TWELVE

Good-Carb Cooking

Now that you know good carbs will get you on the road to good health, what's the next step? Start cooking with them and incorporating them into your daily diet. With so many good carbs available, you can have a field day in your kitchen. What you will find here are thirty-two good-carb recipes for every possible course, from dips to desserts. Each of these recipes has been tested by the organizations that created them and use many of the very best carbs discussed in this book. *Bon appetit!*

Dips

Avo Salsa
Serves 8

2 ripe medium California avocados, peeled, pitted, and diced

1 large ripe tomato, diced

¼ cup finely chopped red onion

2 cloves garlic, minced

2 tablespoons chopped fresh cilantro

Juice of 1 large lime

½ teaspoon ground cumin

½ teaspoon freshly ground black pepper

½ teaspoon salt

1. In a medium bowl, combine avocados, tomato, onion, garlic, cilantro, lime juice, cumin, pepper and salt.

2. Toss well and maintain chunky consistency.

Reprinted with permission from the California Avocado Commission, www.avocado.org.

California Avocado Green Onion Dip
Serves 7

1 medium California avocado, seeded and peeled

1 tablespoon fresh lemon juice

1 cup 1 percent low-fat cottage cheese

¾ cup plain nonfat yogurt

¼ cup nonfat mayonnaise

4 green onions, thinly sliced (about ½ cup)

¼ cup carrots, shredded (about ½ medium carrot)

1½ cups broccoli florets

1 cup cucumber slices

28 melba toast rounds

1. Dice avocado into small pieces, toss with lemon juice, and set aside.

2. In a food processor or blender, blend cottage cheese, yogurt, and mayonnaise until smooth.

3. Add cottage cheese mixture to avocado, gently stirring in onions and carrots.

4. Cover and chill. Serve with broccoli, cucumber, and melba toast, allowing ½ cup vegetables, 4 melba toast rounds, and 8 tablespoons dip per serving.

Reprinted with permission from the California Avocado Commission, www.avocado.org.

Artichokes with Light Honey-Mustard Dip
Makes about ½ cup

> 4 medium California artichokes
> ½ cup light mayonnaise
> 2 teaspoons honey
> ½ teaspoon mustard
> ½ teaspoon fresh lemon juice

1. Wash artichokes under cold running water. Cut off stem at base; remove small bottom leaves. If desired, trim tips of leaves and cut off top 2 inches. Stand artichokes upright in a deep, nonreactive saucepan large enough to hold snugly.

2. Add 1 teaspoon salt and 2 to 3 inches boiling water. Cover and boil gently 35 to 45 minutes or until artichoke base can be pierced easily with a fork. (Add a little more boiling water, if needed.) Turn artichokes upside down to drain. Serve immediately or cool completely; cover and refrigerate to chill. Makes 4 artichokes.

3. Combine light mayonnaise, honey, mustard, and lemon juice; mix well.

Reprinted with permission from the California Artichoke Advisory Board, www.artichokes. org.

Soups

Boston Bean Soup
Makes 4 servings

> 1 cup dried pinto beans, or 2 (15-ounce) cans cooked pinto beans, drained
> 2 medium tomatoes, seeded and chopped
> 1 rib celery, sliced
> 1 medium onion, chopped
> 1 bay leaf
> 1 (15-ounce) can reduced-sodium, fat-free beef broth
> Salt and freshly ground black pepper to taste

1. If using canned beans, ignore this step. If using dried beans, place in a small Dutch casserole. Add 3 cups cold water. Cover and bring to a boil. Remove from heat and soak one hour. Drain well.

2. In a medium pan, mix together beans, tomatoes, celery, onion, bay leaf, and broth. Cover and bring to boil over medium-high heat. Reduce heat and simmer until vegetables are quite soft, 60 to 75 minutes for dried beans, 20 minutes for canned. Let hot soup sit, uncovered, 20 minutes. Remove bay leaf.

3. Purée half the soup in a blender. Recombine with remaining soup. Season to taste with salt and pepper.

Reprinted with permission from the American Institute of Cancer Research, www.aicr.org.

Kidney Bean and Quinoa Chowder

Makes 4 servings

2 teaspoons olive oil

¾ cup chopped onion

1 rib celery, cut in ½-inch slices

¼ cup quinoa, rinsed well and drained

1 small zucchini, cut in ½-inch cubes

1 medium red or white potato, cut in ½-inch cubes

½ Granny Smith or Fuji apple, peeled, cored, and cut in ½-inch cubes

6 cups vegetable broth

1 (10-ounce) can kidney beans, drained and rinsed

1 cup fresh, frozen, or canned corn kernels

Salt and freshly ground pepper to taste

2 tablespoons chopped fresh cilantro, for garnish

1. Heat oil in a large saucepan over medium-high heat. Sauté onion and celery until onion is translucent, about 5 minutes.

2. Add quinoa, zucchini, potato, and apple. Pour in broth. Bring to a boil, reduce heat, and simmer until potatoes are tender and grain is cooked, about 15 minutes.

3. Add beans and corn. Cook until heated through. Season chowder to taste with salt and pepper. Ladle into warm bowls, garnish with cilantro, and serve.

Variation: Along with beans and corn, add 1 to 2 corn tortillas, torn into bite-size pieces, or ¾ cup cubed cooked chicken.

Cooking for two: This soup keeps two to three days in the refrigerator and reheats well. So take a break, then serve it again in a couple days. The second time, garnish with a squirt of fresh lime juice instead of cilantro.

Reprinted with permission from the American Institute of Cancer Research, www.aicr.org.

Salads

Spinach Salad

1 pound spinach, torn into bite-size pieces
1 medium red onion, thinly sliced
1 small can mandarin oranges, drained
½ cup almonds

Combine spinach, onion, oranges, and almonds and serve with desired dressing.

Reprinted with permission from the Pioneer Valley Growers Association, www.pvga.net.

Sweet Potato Salad

Makes 6 servings

3 pounds sweet potatoes
1½ cups nonfat plain yogurt
2 tablespoons fresh cilantro, minced
2 tablespoons shallots, minced
1 tablespoon fresh lime juice
Salt and freshly ground black pepper to taste
1 green bell pepper, seeded and chopped

2 celery stalks, chopped

Chopped canned chiles to taste

1. Preheat oven to 400 degrees.

2. Scrub sweet potatoes and pierce all over with a fork. Bake until soft, about 1 hour. While potatoes are baking, prepare dressing.

3. Make dressing by mixing together yogurt, cilantro, shallots, and lime juice. Add salt to taste. Chill at least 1 hour before using to dress salad.

4. When sweet potatoes are baked, cool, peel, and cut in 1/2-inch cubes. Place in large bowl and mix in salt and pepper to taste. Mix potatoes with bell pepper, celery, and chiles.

5. Mix dressing into potato salad. Serve warm or chilled.

Reprinted with permission from the American Institute of Cancer Research, www.aicr.org.

Curried Couscous Salad

Makes 6 servings

1¼ cups fat-free, reduced-sodium chicken broth

1 tablespoon curry powder

1 tablespoon extra-virgin olive oil, divided

¾ cup couscous

½ cup carrot, cut in ½-inch dice

½ cup Spanish onion, cut in ½-inch dice

½ cup tomato, seeded and cut in ½-inch dice

½ cup zucchini, cut in ½-inch dice

¼ cup dried currants

2 tablespoons fresh lemon juice

Salt and freshly ground pepper to taste

1. In a medium saucepan over medium-high heat, combine broth, curry powder, and 2 teaspoons oil. Bring to a boil. Stir in couscous, reduce heat, cover, and cook 1 minute. Remove from heat and let couscous sit, covered, 10 minutes.

2. Fluff couscous with fork and pour into a large bowl. Stir in carrot, onion, tomato, zucchini, and currants.

3. In a small bowl, combine lemon juice, salt, and pepper. Whisk in remaining 1 teaspoon oil. Pour dressing over salad. Toss with fork until all ingredients are combined. Season to taste with salt. Serve warm or at room temperature.

Variation: Mix in ½ cup canned chickpeas or ½ cup shredded, cooked chicken breast.

Reprinted with permission from the American Institute of Cancer Research, www.aicr.org.

Golden Fruit Salad
Makes 6 to 8 servings

> 1 medium mango
> 1 Gala or Golden Delicious apple, peeled, cored, and thinly sliced
> ½ Asian pear, peeled, cored, and thinly sliced
> 1 peach or nectarine, thinly sliced
> ½ cup red seedless grapes, halved
> 6 whole dried apricots, or 9 halves, cut in ½-inch slivers
> 1 tablespoon finely chopped, candied or preserved ginger
> ½ cup orange juice
> ½ teaspoon vanilla

1. To cube mango, place on counter. Holding a knife horizontally, cut off one side of mango, slicing as close to pit as possible. Turn fruit over and repeat to remove other side. Hold one half in the palm of your hand, skin-side down. Using top of knife, score mango vertically and horizontally, making cuts about ¾ inch apart and slicing fruit so you feel tip of knife against skin of fruit. Grasp two opposite sides of fruit between your thumb and fingers and turn skin back so scored squares stand out like a porcupine's quills. Holding the knife horizontally, carefully cut cubes of mango at their base, separating them from skin. Place cubed fruit in a large serving bowl.

2. Add apple, Asian pear, peach or nectarine, grapes, and apricots to mango and toss gently to combine. Add ginger, orange juice, and vanilla, then toss again.

3. Let fruit salad sit 15 minutes, at room temperature, so flavors can meld.

Serve immediately or cover with plastic and refrigerate 3 to 4 hours. The fruit becomes mushy if left longer.

Reprinted with permission from the American Institute of Cancer Research, www.aicr.org.

Marinated Vegetable Salad

Makes 8 servings

1 (16-ounce) package frozen California blend vegetables
½ cup sugar
1 tablespoon flour
1 teaspoon dry mustard
½ teaspoon salt
½ cup vinegar
1 medium onion, chopped
1 (15 1½-ounce) can dark red kidney beans, drained and rinsed
½ cup celery, chopped
1 green pepper, chopped

1. Cook frozen vegetables according to package directions. Cool.

2. For dressing: Combine sugar, flour, dry mustard, and salt. Add vinegar. Cook until clear, stirring constantly. Allow dressing to cool.

3. For salad: Combine cooked vegetables, celery, green pepper, onion, and beans. Add dressing and toss to mix. Refrigerate several hours to blend the flavors. Stir beans into salad just before serving.

Reprinted with permission from the Northarvest Bean Growers Association, www.northarvest bean.org.

Lemon Blueberry and Chicken Salad

Makes 4 servings

¾ cup low-fat lemon yogurt
3 tablespoons reduced-calorie mayonnaise
1 teaspoon salt

    **2 cups fresh or frozen blueberries, divided (Reserve a few blueberries
for garnish.)**
    2 cups cubed cooked chicken breasts
    ½ cup sliced green onions (scallions)
    ¾ cup diagonally sliced celery
    ½ cup diced sweet red bell pepper

1. In a medium bowl, combine yogurt, mayonnaise, and salt. Add blueberries, chicken, green onions, celery, and bell pepper; mix gently. Cover and refrigerate to let flavors blend, at least 30 minutes.

2. Serve over endive or other greens and garnish with reserved blueberries and lemon slices, if desired.

Reprinted with permission from the U.S. Highbush Blueberry Council, www.blueberry.org.

Blueberry Balsamic Vinegar

Makes 5½ cups

    4 cups frozen, thawed, or fresh blueberries
    1 quart balsamic vinegar
    ¼ cup sugar
    Lime peel cut in strips from 1 lime (green part only)
    1 (3-inch) cinnamon stick

1. In a large nonreactive saucepan, crush blueberries with a potato masher or back of a heavy spoon. Add vinegar, sugar, lime peel, and cinnamon stick; bring to a boil. Reduce heat and simmer, covered, for 20 minutes. Cool slightly and pour into a large bowl. Cover and refrigerate for 2 days to allow flavors to blend.

2. Place a wire mesh strainer over a large bowl. In batches, ladle blueberry mixture into strainer, pressing out as much liquid as possible. Discard solids.

3. Pour vinegar into clean glass bottles or jars; refrigerate, tightly covered, indefinitely. Use this in salad dressings or drizzled over grilled chicken or beef.

Reprinted with permission from the U.S. Highbush Blueberry Council, www.blueberry.org

Greek Potato Salad with Dried Tomatoes

Makes 4 servings

1 pound (3 medium) potatoes, uniform in size, cut into ¼-inch slices

Lemon Dressing

¼ cup olive oil
¼ cup water
2 ½ tablespoons lemon juice
1 large clove garlic, pressed
1 tablespoon chopped fresh oregano, or 1 teaspoon dried oregano leaves
1 teaspoon salt
½ teaspoon pepper
1 cup (1½ ounces) dried tomato halves, halved with kitchen shears
1 cup sliced seedless cucumber
½ cup sliced red onion
1 cup crumbled feta cheese
½ cup Greek olives or pitted ripe olives

1. In a 2-quart saucepan over medium heat, cook potatoes, covered, in 2 inches boiling water until tender, about 12 minutes; drain and set aside.

2. Meanwhile, in a small bowl, cover tomatoes with boiling water; set aside 10 minutes while you whisk together olive oil, water, lemon juice, garlic, oregano, salt, and pepper for dressing.

3. Thoroughly drain tomatoes and pat dry with paper towels. Add potatoes, tomatoes, and cucumbers to bowl containing dressing; toss to coat. Mound potato mixture on plate. Arrange onion, cheese, and olives on top.

Reprinted with permission from the U.S. Potato Board, www.potatohelp.com.

California Bulgur Salad with Lemon Mint Dressing

Makes 4 servings
Dressing makes about ½ cup

1 cup bulgur wheat
½ cup sliced mushrooms
2 tablespoons butter or margarine

2 cups chicken broth

¼ cup diagonally sliced green onions

Lemon Mint Dressing (recipe follows)

1 cup finely shredded red cabbage

2 to 3 (about 3 to 3½ ounce each) California kiwifruit, pared and sliced

1 (10- to 12-ounce) cooked chicken breast boned, skinned, and sliced

1. Sauté bulgur and mushrooms in butter until golden; add chicken broth, cover, and bring to boil. Reduce heat and simmer 15 minutes. Cool. Toss with green onions and ¼ cup Lemon Mint Dressing. Arrange bulgur, cabbage, and kiwifruit on plates with sliced chicken or fish; drizzle with remaining Lemon Mint Dressing.

2. Lemon Mint Dressing: Combine 6 tablespoons lemon juice, 2 tablespoons vegetable oil, and 4 teaspoons honey with 2 teaspoons each grated lemon peel and fresh minced mint leaves (one teaspoon dried crushed mint can be substituted); mix well.

Reprinted with permission from the California Kiwifruit Commission, www.kiwifruit.org

Main Dishes

Black Bean Burgers
Makes 4 servings

Canola oil cooking spray

2 bunches finely chopped scallions, both white and green parts

1 red pepper, seeded and cut in ½-inch pieces

2 cloves garlic, finely chopped

1 (15-ounce) can black beans, drained and rinsed

1 cup cooked brown rice

Dash hot pepper sauce, or to taste

1 teaspoon cumin, or to taste

Salt and freshly ground black pepper to taste

1 large egg white, lightly beaten

½ cup whole-grain breadcrumbs

1. Heavily coat a medium skillet with canola oil cooking spray. Heat over medium-high heat until hot. Add scallions, red pepper, and garlic. Reduce heat to medium-low and sauté until very soft, about 5 minutes. Do not let vegetables brown.

2. Remove vegetables from heat and mix in beans and rice. Transfer to a food processor or blender and process until mixture is coarsely chopped. Be careful not to overprocess.

3. Transfer mixture to a medium bowl. Season to taste with hot pepper sauce, cumin, salt, and pepper. Add egg white and mix in lightly with a fork until just blended. Mix in breadcrumbs with a fork until lightly blended. Form mixture into eight patties. (Patties will hold their shape better if refrigerated, covered, at least 30 minutes.)

4. When ready to sauté patties, lightly coat a skillet with cooking oil spray and heat over medium-high heat until hot. Add patties and sauté on both sides until nicely browned, about 4 minutes per side.

5. Serve plain or with lettuce and tomato on whole-grain buns.

Reprinted with permission from the American Institute of Cancer Research, www.aicr.org

Roasted Vegetable Wrap

Makes 4 servings

> 1 small eggplant
> 3 teaspoons extra-virgin olive oil, divided
> ½ small bulb fennel, cut vertically in very thin slices
> 2 large mushrooms, thinly sliced
> 1 small zucchini, thinly sliced
> 1 small red pepper, seeded and cut in ¼-inch strips
> 8 thin slices red onion
> 8 to 12 whole garlic cloves, peeled
> 2 teaspoons minced fresh rosemary, or ½ teaspoon dried and crushed
> Salt and freshly ground pepper to taste
> 2 tablespoons soft, fresh goat cheese
> 1 (15-inch) piece soft cracker bread, or 2 (9-inch) wheat tortillas

1. Preheat oven to 375 degrees. Spray two baking sheets with cooking spray.

2. Cut 4 slices vertically from eggplant, each ¼-inch thick. Set aside remaining eggplant for another use. Lay slices in one layer on one of the baking sheets. Pour ½ teaspoon olive oil on your palm and rub it over slices, turning so both sides are lightly coated. Do the same with fennel slices. Arrange both vegetables on the baking sheet. Roast until eggplant is tender, 12 to 15 minutes.

3. Meanwhile, toss mushrooms and zucchini in a bowl with 1 teaspoon olive oil. Arrange vegetables in one layer on second baking sheet. Roast until tender, about 10 minutes. Set aside.

4. Toss red pepper, onion, and garlic with remaining oil, rosemary, and, if desired, salt and pepper. Reusing the baking sheet from eggplant, spread vegetables in one layer and roast until softened, about 15 minutes. Set aside.

5. To assemble wrap, thinly spread goat cheese on one side of bread to cover it completely. Arrange vegetables to cover two-thirds of bread, keeping cheese-only part at top. Start with eggplant, followed by fennel, red pepper mixture, and mushroom-zucchini mixture.

6. Roll up filled bread, jelly-roll fashion, starting at the bottom. To keep filling from pushing forward, keep pulling rolled part toward you. This also helps make a firm roll. Wrap and refrigerate from 4 to 48 hours. To serve, cut into 2-inch slices.

Variation: Use whole-wheat pita bread in place of cracker bread or tortillas.

Cooking for two: Make all the vegetables. Use the extra as a side dish over the next day or two.

Reprinted with permission from the American Institute of Cancer Research, www.aicr.org.

Speedy Meatless Taco Filling
Makes 4 servings

1 tablespoon canola oil
1 medium onion, chopped
2 garlic cloves, finely chopped
1 medium green bell pepper, seeded and chopped
1 teaspoon ground cumin

1 teaspoon dried oregano

½ cup fresh cilantro leaves, chopped

1 medium tomato, seeded and chopped

1 (12-ounce) package refrigerated or frozen soy crumbles

1 ½ cups prepared salsa

Dash of Tabasco sauce, or to taste (optional)

Salt and freshly ground black pepper to taste

1. In a large, nonstick skillet, heat oil until hot. Sauté onion, garlic, and green pepper until onion is translucent, about 4 minutes. Add cumin and oregano and mix until fragrant.

2. Add cilantro, tomato, soy crumbles, and salsa. Bring to a boil, reduce heat, and simmer 3 minutes. Season to taste with Tabasco sauce, salt, and pepper. Serve in taco shells, over cooked brown rice, or use to make nachos. Can store refrigerated up to 4 days.

Reprinted with permission from the American Institute of Cancer Research, www.aicr.org.

White Chili

Makes 12 servings

1 pound dry navy beans

1 pound ground turkey, or 2 cups cooked, cubed turkey

2 (4-ounce) cans medium or hot green chili peppers, diced

3 tablespoons dry chicken bouillon

2 medium onions, chopped

2 teaspoons dry minced garlic

2 tablespoons whole cumin

4 ounces low-fat Monterey Jack cheese, shredded

1. Soak beans using Preferred Hot Soak method: Add 10 cups of cold water to the beans. Bring the water to a boil, and boil for 3 minutes. Cover the pot, letting the beans soak for at least 4 hours. Then, drain and rinse soaked beans. Add fresh cold water to fully cover beans. Simmer the beans until they're tender. Brown ground turkey, drain, and discard fat. Combine in a slow cooker: beans and liquid, turkey, 1 can chili peppers, bouillon, onions, garlic, and cumin. Stir.

2. Cover and cook on low for several hours. Taste and add more chili peppers 1 tablespoon at a time if a hotter taste is desired. (Freeze leftover chili peppers.) Serve hot, topped with cheese.

Reprinted with permission from the Northarvest Bean Growers Association, www.northar vestbean.org.

Side Dishes

Red Potatoes with Kale
Makes 4 servings

> 4 medium red potatoes
> 1 bunch kale
> 1 teaspoon toasted sesame oil
> 1 onion, thinly sliced
> 2 garlic cloves, minced
> ½ teaspoon black pepper
> ½ teaspoon paprika
> 5 teaspoons soy sauce

1. Scrub potatoes and cut into ½-inch cubes or wedges. Steam over boiling water until just tender when pierced with fork. Rinse with cold water, drain, and set aside.

2. Rinse kale and remove tough stems. Tear leaves into small pieces.

3. Heat oil in a large nonstick skillet and add onion and garlic. Sauté for 5 minutes.

4. Add cooked potatoes, pepper, and paprika, and continue cooking until potatoes begin to brown, about 5 minutes. Turn mixture gently as it cooks.

5. Spread kale leaves over top of potato mixture. Sprinkle with 2 tablespoons water and soy sauce. Cover and cook, turning occasionally, until kale is tender, about 7 minutes.

Reprinted with permission from the American Institute of Cancer Research, www.aicr.org.

Pineapple, Corn, and Mango Salsa

Makes 2½ cups

> 1 cup canned crushed pineapple (packed in its own juices), drained
> ½ medium mango, diced
> ½ cup frozen corn, thawed
> ½ cup chopped tomatoes
> ¼ cup minced parsley
> 3 tablespoons minced red onion
> Salt, cayenne pepper, and cumin to taste

1. In medium bowl, mix pineapple, mango, corn, tomatoes, parsley, and onion. Season with salt, cayenne, and cumin to taste.

2. Serve over grilled fish, chicken, or tofu.

Reprinted with permission from the American Institute of Cancer Research, www.aicr.org.

Risotto Primavera

Makes 5 servings

> 3 cups fat-free, reduced-sodium chicken stock or broth
> 1 small green zucchini squash, cut in ½-inch dice
> 6 thin asparagus stalks, cut in ½-inch pieces, tips reserved
> 1 medium carrot, halved lengthwise and thinly sliced
> 1 tablespoon extra-virgin olive oil
> ¼ cup finely chopped Spanish onion
> 1 cup Arborio rice
> 2 teaspoons lemon juice, preferably fresh
> 1 small garlic clove, minced
> ½ cup fresh or frozen baby green peas
> ¼ cup chopped fresh flat-leaf parsley
> 1 tablespoon low-fat yogurt
> 2 tablespoons grated Parmigiano-Reggiano cheese
> Salt and freshly ground black pepper to taste

1. Heat chicken stock to boiling. Set aside.

2. Place zucchini in a large bowl. Add asparagus and carrot and mix in.

3. Heat oil in a deep saucepan over medium-high heat. Add onion and sauté until translucent, about 2 minutes. Mix in rice until coated with oil and opaque, about 1 minute. Add lemon juice, stirring until rice is almost dry, less than 1 minute. Mix in garlic and half the chopped vegetables. Cook 1 minute.

4. Add hot broth, $\frac{1}{2}$ cup at a time, stirring well after each addition. Cook, stirring continually, until rice is almost dry before adding more broth. When most of broth has been used and rice is almost done but has a hard core, about 15 to 18 minutes, add remaining vegetables and parsley. Add remaining broth and cook until rice is tender but still al dente (offering a slight resistance when bitten into, but not soft), about 3 to 4 minutes.

5. Remove pot from heat. Stir in yogurt and cheese. Season to taste with salt and pepper. Serve immediately.

Reprinted with permission from the American Institute of Cancer Research, www.aicr.org

Grilled Potato Planks

Makes 4 servings

> 3 tablespoons olive oil
> 1 clove garlic, minced
> 2 teaspoons finely chopped fresh rosemary leaves
> $\frac{1}{2}$ teaspoon salt
> 1½ pounds (about 3 large) unpeeled baking potatoes, cut into
>   ½ inch-thick slices

1. Preheat grill. Combine oil, garlic, rosemary, and salt in dish. Add potato slices and turn until well coated. Grill potatoes for 8 minutes or until soft. Turn and continue grilling 10 minutes longer or until cooked through.

2. Remove from grill and serve with your favorite grilled meals.

Reprinted with permission from the U.S. Potato Board, www.potatohelp.com.

Garlic Roasted Potatoes

Makes 2 servings

¾ pound (2 medium) potatoes, cut into wedges

1 tablespoon olive oil

2 small cloves garlic, finely chopped

½ teaspoon salt

¼ teaspoon pepper

1½ teaspoons finely chopped fresh parsley

1. Preheat oven 400 degrees. In a large bowl, toss together potatoes, oil, garlic, salt, and pepper until potatoes are well coated. Arrange potatoes in a single layer on a large baking sheet. Bake 1 hour or until browned and crisp, turning potatoes twice with a spatula during cooking.

2. Remove potatoes from oven and toss with parsley.

Reprinted with permission from the U.S. Potato Board, www.potatohelp.com.

Crunchy Seasoned Oven Fries

Makes 4 servings

1 egg

1⅓ pounds (4 medium) potatoes, cut into ½-inch-thick wedges

¾ cup cornflake crumbs

Italian Fries Seasoning

6 tablespoons grated Parmesan cheese

1½ teaspoons Italian herb seasoning

Chili Fries Seasoning

1½ tablespoons chili powder

1½ teaspoons garlic salt

1. Heat oven to 375 degrees. Coat two baking sheets with vegetable cooking spray. Lightly beat egg in shallow bowl. In another shallow bowl, mix crumbs and seasoning blend of your choice. Dip potato wedges into egg, then coat completely with crumb mixture. Arrange in a single layer on baking sheets. Bake 20 minutes, then turn potatoes over and continue to bake 10 to 15 minutes longer, until potatoes are

browned and crisp and insides are tender when tested with toothpick or fork.

2. Serve immediately, plain or with lemon or ketchup.

Reprinted with permission from the U.S. Potato Board, www.potatohelp.com

Grilled Artichokes
Makes 8 servings—½ artichoke each

> **4 large artichokes**
> **¼ cup balsamic vinegar**
> **¼ cup water**
> **¼ cup soy sauce**
> **1 tablespoon minced ginger**
> **¼ cup olive oil**

1. Slice artichoke tops off, crosswise. Trim stems.

2. Boil or steam artichokes until bottoms pierce easily or a petal pulls off easily.

3. Drain artichokes. Cool. Cut each artichoke in half lengthwise and scrape out fuzzy center and any purple-tipped petals.

4. Mix vinegar, water, soy sauce, ginger, and olive oil in a large plastic bag. Place artichokes in the bag and coat all sides of artichokes. For best flavor, marinate overnight in the refrigerator, but should marinate at least 1 hour.

5. Drain artichokes. Place cut side down on a grill over a solid bed of medium coals or gas grill on medium. Grill until lightly browned on the cut side, 5 to 7 minutes. Turn artichokes over and drizzle some of remaining marinade over artichokes. Grill until petal tips are lightly charred, 3 to 4 minutes more.

6. Serve hot or room temperature.

Reprinted with permission from the California Artichoke Advisory Board, www.artichokes.org.

Butternut Squash with Ginger

Makes 4 servings

> 1 large butternut squash
> 1 tablespoon ginger root, freshly minced
> 1/4 cup unsweetened apple juice
> Nutmeg, freshly ground

1. Preheat oven to 350 degrees.

2. Peel and seed squash and cut into 1/2-inch cubes. Put squash, ginger root, and apple juice into a lightly oiled baking dish.

3. Cover and bake for 50 to 60 minutes. Sprinkle with nutmeg just before serving.

Reprinted with permission from the Pioneer Valley Growers Association, www.pvga.net.

Beet and Tomato Casserole

Makes 6 servings

> 2 1/2 cups canned beets, sliced
> 2 1/2 cups canned tomatoes
> 1/2 cup grated cheese, any type
> Salt and pepper to taste
> 2 cups breadcrumbs
> 1 tablespoon butter

1. Preheat oven to 350 degrees.

2. Put half the beets in the bottom of a greased baking dish. Add half the tomatoes then half the cheese in layers. Add salt and pepper, if desired. Add half the breadcrumbs. Dot with 1 tablespoon butter. Repeat with the rest of ingredients.

3. Cook for 20 minutes until brown.

Reprinted with permission from the Pioneer Valley Growers Association, www.pvga.net

Desserts

Red Berry Kissel
Makes 4 servings

   ½ (20-ounce) bag whole, unsweetened frozen strawberries
     (about 10 ounces)
   1 (10-ounce) package frozen sweetened raspberries
   ½ cup cranberry cocktail juice
   3 tablespoons cornstarch
   3 tablespoons cold water
   ½ teaspoon almond extract

1. Place frozen strawberries and raspberries in a deep saucepan. Add cranberry juice. Over medium-high heat, bring just to boil. Reduce heat and simmer until berries are very soft, about 20 minutes.

2. Pour berry mixture into fine sieve held over a bowl. With a wooden spoon, push berry pulp through the sieve. Scrape strained berries on the outside of the sieve into the bowl.

3. Rinse out and dry pot. Rinse four dessert dishes in cold water, but do not dry. Set aside.

4. Whisk berry mixture to combine pulp and liquid well. Return mixture to pot. Mix cornstarch and water in a small bowl. Stir mixture into berries. Add almond extract.

5. Over medium heat, cook mixture until translucent, stirring constantly. When mixture heavily coats spoon and thickens—about 1 to 2 minutes—remove from heat before it comes to a boil and pour into dessert dishes. When almost cool, refrigerate. To prevent surface skin from forming, cover bowls with plastic wrap, pressing it to touch the surface of pudding. Kissel can be made up to 2 days ahead.

6. Let kissel sit 20 minutes at room temperature before serving.

Reprinted with permission from the American Institute of Cancer Research, www.aicr.org.

Great Grilled Fruit Kebabs

Makes 8 servings

> 2 tablespoons canola oil
> 2 tablespoons brown sugar
> 2 tablespoons fresh lemon juice
> 1 teaspoon cinnamon
> 4 (1-inch) slices pineapple, canned or fresh, cut into chunks
> 2 apples, cored and cut into 1-inch pieces
> 2 pears, pitted and cut into 1-inch pieces
> 2 peaches, nectarines, or plums (or a mix), pitted and cut into
>   1-inch pieces
> 2 bananas, peeled and cut into 1-inch pieces

1. In a small bowl, stir together oil, brown sugar, lemon juice, and cinnamon until sugar is dissolved.

2. Thread pineapple, apples, pears, peaches, and bananas alternately onto each of eight skewers. Brush kebabs with oil mixture and place skewers on barbecue grill. Turn frequently until fruit starts to brown, about 6 to 8 minutes.

Reprinted with permission from the American Institute of Cancer Research, www.aicr.org.

Blueberry Orange Whirl

Makes 4 servings

> 1 (12-ounce) package frozen blueberries, unthawed, or 2½ cups
>   fresh blueberries
> 1 (8-ounce) container vanilla low-fat yogurt
> ½ cup orange juice
> ½ cup milk
> 1 teaspoon vanilla extract

1. In an electric blender, whirl blueberries, yogurt, orange juice, milk, and vanilla extract until smooth.

2. Serve immediately.

Reprinted with permission from the U.S. Highbush Blueberry Council, www.blueberry.org.

Blueberry Granola Bars
Makes 18 bars

   ½ cup honey
   ¼ cup firmly packed brown sugar
   3 tablespoons vegetable oil
   1½ teaspoons ground cinnamon
   3½ cups quick-cooking oats
   2 cups fresh blueberries

1. Preheat oven to 350 degrees. Lightly grease a 9×9-inch square baking pan. In a medium saucepan, combine honey, brown sugar, oil, and cinnamon. Bring to a boil, and boil for 2 minutes; do not stir. In a large mixing bowl, combine oats and blueberries. Stir in honey mixture until thoroughly blended. Spread into prepared pan, gently pressing mixture flat. Bake until lightly browned, about 40 minutes.

2. Cool completely in the pan on a wire rack. Cut into 1½×3-inch bars.

Reprinted with permission from the U.S. Highbush Blueberry Council, www.blueberry.org.

Kiwifruit Frozen Yogurt
Make 6 servings

   2 California kiwifruit, peeled and coarsely chopped
   1 tablespoon honey
   1 pint frozen low-fat or nonfat vanilla yogurt, softened
   1 to 2 drops green food coloring (optional)
   1 (10-ounce) package frozen red raspberries in syrup, thawed
   2 tablespoons triple sec or other orange liqueur
   2 teaspoons cornstarch
   3 California kiwifruit, ends trimmed and sliced lengthwise
   Fresh mint leaves
   Fresh or frozen whole raspberries (optional)

1. In a food processor or blender, purée 2 chopped kiwifruit; stir in honey. Place in freezer and freeze until slushy (about 45 minutes). In a stainless-steel bowl, quickly combine softened yogurt, kiwifruit mixture, and food coloring (if using); refreeze in a bowl. With a small ice-cream scoop

(about 2 tablespoons), form 12 balls and place on wax paper-lined tray; refreeze.

2. Meanwhile, to make sauce, in food processor or blender, purée thawed raspberries. Over a saucepan, strain berries through a fine sieve, pressing with the back of a spoon. Discard seeds. Stir in triple sec and cornstarch. Bring to boil, stirring constantly until slightly thickened. Cool; cover and chill. To assemble, spoon about 2 tablespoons sauce on each of six dessert plates or shallow bowls. Arrange kiwifruit slices and frozen yogurt balls on sauce. Garnish with mint leaves and whole raspberries.

Reprinted with permission from the California Kiwifruit Commission, www.kiwifruit.com.

GOOD FAT
vs.
BAD FAT

To my brother Tom and my sister Gretchen, with love

• ACKNOWLEDGMENTS •

I gratefully thank the following people for their work and contributions to this book: my agent Madeleine Morel, 2M Communications, Ltd.; Christine Zika and the staff at The Berkley Publishing Group; and my husband, Jeff, for love and patience during the research and writing of this book.

Part I

Fats: The Good, the Bad, and the Ugly

The Fats of Life

Fats

They're the black-hatted bad guys of nutrition, the most demonized of all nutrients. Even the word *fat* itself conjures up all sorts of negative associations when used in everyday language. Take the expressions *fathead, fat chance, fat cat,* and *fatso,* for example. *Fat* is practically a four-letter word.

But is fat really all that bad? Does it deserve to be so maligned?

This may come as a surprise to you, but fats are mostly good guys in nutrition. You need them to survive. In fact, there are a slew of "good" fats with astonishing powers to outwit disease and keep you healthy for a lifetime. Sure, there are some health-risky fats, but even some of those are needed in small amounts for good health. When you're dealing with fats, the key is to control not only the amount you eat but also the kind of fat you eat.

To get a handle on how fats affect your health, it helps to learn some basic facts about this most misunderstood of all nutrients. So let's get started on a short nutrition lesson.

Fat Facts

You've heard the old expression, "Oil and water don't mix." Well, fats—many of which are oils—are members of a family of chemical compounds technically known as *lipids* that for the most part don't dissolve in water. You know this if you've ever made salad dressing and watched the fatty part separate from the rest of the liquid and gradually rise to the top.

When we speak of fat in our foods or on our bodies, we're talking about *triglycerides*. Triglycerides make up about 95 percent of dietary fat and 90 percent of body fat. Some triglycerides also circulate in your bloodstream. Chemically, a triglyceride is a backbone of glycerol (a type of alcohol) to which three fatty acids are attached, hence the name triglyceride.

A *fatty acid* is a building block of fat. Many specific types of fatty acids are found in various fats, each with different properties that influence your health in far-reaching ways. Fatty acids are contructed of chains of carbon atoms with hydrogen atoms attached, with an acid group at one end. Think of this configuration as a charm bracelet. The carbons form the chain, and the hydrogen and the acid group are the charms.

The lengths of these chains vary according to the fat. Fats found in meat, for example, usually have chains that are sixteen or more carbons long. Some carbon chains are much shorter, with six, eight, ten, or twelve carbon atoms.

Is the length of a fat chain important?

Yes. Here's the deal: Length has a lot to do with how your body uses the fat and obtains energy from it. Short- and medium-chain fatty acids, which are generally found in butter and coconut oil, are a good example. During digestion, they are absorbed more readily by your body than longer-chain fatty acids are and thus supply quick energy. Because of this, short- and medium-chain fatty acids are less likely to be packed away as body fat. Longer-chain fatty acids, on the other hand, tend to be stored as fat. So to a certain extent, length matters.

Saturated and Unsaturated Fatty Acids

Fatty acids from food are chemically classified not only by the length of their chains but also according to the number of hydrogens the fatty acid chain holds. This attribute is referred to as *saturation*.

When a fatty acid carries the maximum number of hydrogen atoms, it is said to be loaded or *saturated*. If there are one or more places in the chain where hydrogens are missing, the fatty acid is *unsaturated*. A fatty acid with a single point of unsaturation is termed *monounsaturated;* a fatty acid with two or more points is called *polyunsaturated*.

The degree of saturation affects the temperature at which the fat melts. Generally speaking, the more saturated the fatty acids of a fat are, the more solid the fat is at room temperature. Examples of saturated fats include those found in beef, butter, lard, and dairy products. Unsaturated fats such as vegetable oils are usually liquid at room temperature. An exception to the rule that saturated fats are more solid than unsaturated fats is coconut oil, a saturated fat that is liquid at room temperature.

Monounsaturated fatty acids, a type of unsaturated fat, are found in such foods as olive oil, olives, avocado, cashew nuts, and cold-water fish such as salmon, mackerel, halibut, and swordfish. The most common monounsaturated fatty acid in our food is oleic acid, a major component of olive oil. Monounsaturated fats are generally liquid at room temperature but will partially solidify when refrigerated.

The other type of unsaturated fat is polyunsaturated fat. Found in fish and in most vegetable oils, these foods are endowed with vitaminlike nutrients known as essential fatty acids (EFAs) that your body needs for normal cell growth and development. EFAs are so important that they deserve further explanation (see the next section).

Some polyunsaturated fats such as shortening and stick margarine are hard at room temperature, but only because they have undergone a process called *hydrogenation,* which solidifies vegetable oils.

All fats, whether of vegetable origin or animal origin, feature some combination of saturated fatty acids, monounsaturated fatty acids, and polyunsaturated fatty acids. Animal fats, for example, contain about 50 percent saturated fat, while most vegetable fats are predominantly polyunsaturated fatty acids.

A myth exists that saturated fats contain more calories than either polyunsaturated or monounsaturated fats do. Not so. All pure fats yield 9 calories per gram, and about 115 to 120 calories per tablespoon. However, there may be a slight variation in calories depending on whether the fat is solid or liquid. Solid fats such as butter or margarine contain some air or water and thus will not have as many calories as the same amount of oil. Reduced-calorie or imitation margarines have even fewer calories because they contain more water.

Essential Fatty Acids

Your body can make nearly all the fatty acids it needs for good health—with the exception of essential fatty acids (EFAs), which must be supplied by your diet. Found in plant oils and fish, EFAs are polyunsaturated fats that regulate an amazing number of cellular processes and are endowed with an impressive list of life-sustaining benefits. More than sixty health problems, from heart disease to inflammatory illnesses to immune disorders, can be treated with EFAs.

Unfortunately, though, about 80 percent of all Americans are deficient in these vital nutrients. A major reason is that most of the fat in our diets has been refined and chemically altered—and, in the process, stripped of its essential fatty acids. Examples of synthetic fats include hydrogenated fats and trans-fatty acids found in shortening, stick margarine, fast foods, and commercially baked products such as crackers and cookies. Trans-fatty acids, in particular, can make up to 5 to 35 percent of the fat in margarines, shortenings, and other hydrogenated fats.

Like two riders trying to hop into the same taxi, synthetic fats compete with essential fatty acids for entry into your metabolic pathways, with processed fats muscling essential fats out of the way most of the time. This metabolic mix-up undermines the healing power of essential fatty acids.

How Do EFAs Work?

In actuality, just two fatty acids are considered "essential": linoleic acid (LA) and alpha-linolenic acid (ALA). Both are required for normal cell structure and function. Basically, they make cell membranes more permeable so that nutrient-carrying fluids can pass into cells and waste materials can leave. By contrast, saturated fats stiffen the membranes of cells, making them impermeable and jeopardizing the health of the cells. Thus, it's vital to consume adequate amounts of linoleic acid and alpha-linolenic acid in your diet every day.

Found mostly in vegetable oils, nuts, seeds, and margarine, linoleic acid helps transport water across the skin and ensures the proper functioning of the pituitary gland, which is involved in growth. These actions make linoleic acid a good treatment for skin problems, as well as a potentially therapeutic agent in growth and development therapies. Available from fish, flaxseed oil, and other vegetable oils, alpha-linolenic acid has been studied for its role in fighting heart disease, stroke, and many other serious illnesses.

Both of these essential fatty acids act as "parents," giving birth to other fatty acids in a process involving enzymes. For instance, alpha-linolenic acid produces two fatty acids that play various health-enhancing roles in the body: eicosapentaenoic acid (EPA) and docosahexaenoic acid (DHA). EPA is a very potent fatty acid that prevents platelets in the blood from abnormal clotting, and it helps reduce inflammation. DHA is an important constituent of the brain and retina. Although your body makes DHA and EPA from alpha-linolenic acid, you can obtain them directly from fish in your diet.

Linoleic acid produces gamma-linolenic acid (GLA), which is converted to dihomo-gamma-linoleic acid (DGLA), and then to arachidonic acid (AA). GLA, in particular, has far-reaching benefits in treating all sorts of diseases, particularly those involving inflammation.

In addition to being synthesized from linoleic acid, arachidonic acid occurs naturally in animal and plant foods. Although necessary for infant brain development, arachidonic acid can be harmful in excessive amounts. Arachidonic acid counters the positive action of eicosapentaenoic acid. For instance, if a platelet has a lot of arachidonic acid in its cell membrane, it will clot more readily. On the other hand, a platelet is less likely to clot if there is a lot of eicosapentaenoic acid in its membrane.

These offspring, or derivative, fatty acids manufacture hormonelike compounds called eicosanoids, which include prostaglandins and leukotrienes. Responsible for many of the healing properties of essential fatty acids, prostaglandins and leukotrienes regulate numerous processes, including blood pressure, normal blood clot formation, blood lipids, immunity, inflammation in response to injury, and many other vital functions.

There are "good" prostaglandins and "bad" prostaglandins. Similarly, there are "less inflammatory" leukotrienes and "pro-inflammatory" leukotrienes. A series of good prostaglandins called prostaglandins 1 (PGE1) are synthesized from DGLA. The job of PGE1 is to reduce inflammation, dilate blood vessels, and inhibit blood clotting.

Another series of good prostaglandins called prostaglandins 3 (PGE3) are derived from EPA. So are the less inflammatory leukotrienes. PGE3 helps your body fight infection.

A group of bad prostaglandins called prostaglandins 2 (PGE2), as well as the pro-inflammatory leukotrienes, is synthesized from arachidonic acid. PGE2 steps up inflammation, constricts blood vessels, and encourages abnormal blood clotting.

It's important to understand these metabolic processes. By altering the type of essential fat you eat, you can manipulate the levels of these eicosanoids in your body to treat inflammation, allergies, high blood pressure, and many other adverse health conditions.

Here's how: It's desirable to reduce levels of arachidonic acid and increase DGLA and EPA. This gives rise to more good prostaglandins and leukotrienes. You can accomplish this goal by cutting back on saturated fats (which encourage the production of arachidonic acid), eating more fish (which is loaded with EPA), and supplementing your diet with vegetable oils high in alpha-linolenic acid such as flaxseed oil.

Generally speaking, your body strives to strike a delicate balance among PGE1, PGE2, PGE3, and the leukotrienes. But unless enough of the good prostaglandins are produced, the bad prostaglandins will gang up on your system and harm your body.

Table 1 illustrates how the essential fatty acids are ultimately converted into prostaglandins and leukotrienes.

What Are Omega Fats?

No doubt, you have heard the term *omega fats* used to describe different types of fats. This is simply another classification of essential fatty acids. The omega-3 fatty acids include alpha-linolenic acid (ALA) and its derivative fatty acids, eicosapentaenoic acid (EPA) and docosahexaenoic acid (DHA). The main omega-6 fatty acids are linoleic acid (LA) and its derivative fatty acids, gamma-linolenic acid (GLA) and arachidonic acid (AA). (The designations 3 and 6 refer to their molecular structures.)

Omega-3 fatty acids are vital for normal growth and development and may play a key role in preventing and treating heart disease, high blood pressure, diabetes, arthritis, and cancer. Omega-6 fatty acids are generally necessary for normal growth, hair and skin health, regulation of metabolism, and reproduction. Omega-3 fatty acids, along with omega-6 fatty acids, are now considered essential fats that must be included in your diet. The major sources of omega fats are listed in Table 2.

There is also an omega-9 fatty acid called oleic acid, a monounsaturated fat found most notably in olive oil. It has a range of talents, particularly in cardiovascular health and cancer protection, although it is not an essential fat. You'll learn more about all of these beneficial fats throughout this book.

Table 1

ESSENTIAL FATTY ACID METABOLISM

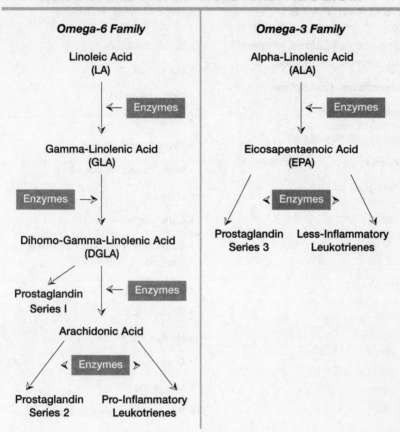

Listed in Table 3 are the various types of dietary fat and their sources, functions, and recommended intake.

Other Forms of Fat

In addition to triglycerides, two other types of lipids are the sterols (which include cholesterol) and phospholipids. Cholesterol is an odorless, waxy, fatlike substance found in all foods of animal origin. It is a chemical cousin of fat. Needed for good health, cholesterol is a constituent of most body tissues and is used to make certain hormones, vitamin D, and bile, a

Table 2

FOODS HIGH IN OMEGA-3 FATS	FOODS HIGH IN OMEGA-6 FATS
• Cold-water fish (salmon, mackerel, tuna, sardines, and herring)	• Black currant oil
• Flaxseeds and flaxseed oil	• Borage oil
• Perilla oil	• Grape-seed oil
• Walnuts and walnut oil	• Sesame oil
• Purslane	• Evening primrose oil
• Wheat germ oil	• Walnuts and walnut oil
	• Corn oil
	• Safflower oil
	• Sunflower seeds and sunflower seed oil
	• Cottonseed oil
	• Soybean oil
	• Hempseed oil
	• Brazil nuts
	• Margarine
	• Pumpkin and squash seeds
	• Spanish peanuts
	• Peanut butter
	• Almonds
	• Wheat germ oil

substance involved in the digestion and absorption of fats. You'll learn more about cholesterol in the next chapter.

Phospholipids contain a molecule of phosphorus, which makes them soluble in water. This characteristic helps fats travel in and out of the lipid-rich membranes of cells. In fact, both cholesterol and phospholipids form part of the structure of cell membranes. A well-known phospholipid is lecithin, manufactured by your liver and found in many foods.

What Happens to the Fat You Eat?

When you feast on fat or fat-containing foods, it begins its digestion in your mouth, where special enzymes act on its breakdown. Fat then travels to your stomach, separates from other food components, and floats to the top of the stomach. Little fat digestion takes place here, however, since fat doesn't mix well with the watery fluids in your stomach.

Fat next enters your small intestine. To assist with digestion here, your gallbladder squirts bile into your intestine at mealtimes. In a process called emulsification, a molecule of bile attaches itself to a molecule of fat, dispersing the fat into the watery solution where it can meet fat-splitting enzymes. Almost the same thing happens when you wash clothes. The detergent acts as an emulsifier to dissolve the grease, molecule by molecule, suspending it in water so that it can be rinsed away. Long-chain fatty acids require bile for digestion; short- and medium-chain acids do not.

If you've had your gallbladder removed, you can still digest fats. That's because the gallbladder only *stores* bile. Bile is produced by your liver and delivered continuously into your small intestine, not just at meals. You'll be instructed to reduce your fat intake, however, because your body can handle only a small amount of fat at a time.

Enzymes split the triglycerides from their glycerol backbones, liberating the fatty acids, which then cross the membrane of the intestine where they are resynthesized back into fats. Some newly re-formed fats are so large that they must be wrapped in special protein blankets to get to their destinations. These protein blankets are called lipoproteins. Some lipoproteins carry a large amount of cholesterol. Elevated concentrations of these lipoproteins can be an early warning sign of heart disease.

Lipoproteins are picked up by the lymphatic system, a secondary circulatory system. It features an elaborate network of organs, tissues, and vessels whose major functions are to transport digested fat from the intestine to the bloodstream and to defend the body against invasion by disease-causing agents.

From the lymphatic system, lipoproteins are carried to the liver (the primary site of metabolism) for further processing. From there, the fat is released into the bloodstream, where it is picked up by fat cells and eventually stored as body fat, if not used immediately for energy. But because the body prefers to use glucose (blood sugar) first for fuel, fats (specifically longer-chain fats) tend to be stored as body fat. By contrast, short-chain fatty acids

Table 3

THE SKINNY ON FATS

Dietary Fats	Sources	Function	Recommended Intake
Saturated fat	Red meat, dairy products, coconut oil, palm oil, egg yolks	Provides energy, stimulates the liver to manufacture cholesterol	7 to 10 percent of total calories
Polyunsaturated fat	Corn, soybean, sesame, safflower oils; some fish	Provides energy, stimulates less bad cholesterol and more good cholesterol	Up to 10 percent of total calories
Monounsaturated fat	Olive, canola, peanut oils; avocado, some fish	Provides energy, stimulates less bad cholesterol and more good cholesterol	Up to 15 percent of total calories
Omega-3 fatty acids	Fish, including salmon, tuna, sardines, bluefish, trout; flaxseed oil	Polyunsaturated fats that reduce the risk of abnormal blood clotting; may help prevent heart disease	Increase as part of total recommended intake of polyunsaturated fats. Optimal ratio of omega-3 fats to omega-6 fats is 1:1; 1:4 is also considered healthy
Omega-6 fatty acids	Many vegetable oils, seeds, nuts, whole grains, fast foods, baked goods	Polyunsaturated fats that are involved in cellular health	Optimal ratio of omega-3 fats to omega-6 fats is 1:1; 1:4 is also considered healthy

Table 3 (continued)

THE SKINNY ON FATS

Dietary Fats	Sources	Function	Recommended Intake
Trans-fatty acids	Stick margarine, shortening, baked goods, fast foods, snack foods	Polyunsaturated fats that have been processed; they act like saturated fats in the body and are damaging to health	Limit or avoid

bypass the lymphatic system and are absorbed directly through a special vein that leads to the liver, where, inside the cells, they are rapidly oxidized or burned up. As you can tell, many complex reactions are involved in the digestion and absorption of dietary fats.

Fat—Who Needs It?

You do! Despite its bad rap, fat is a highly useful nutrient, vital for life. As noted earlier, dietary fat provides essential fatty acids, which are vitaminlike substances that have a protective effect on your body. And once digested, dietary fats help transport and distribute fat-soluble vitamins (vitamins A, D, E, and K) throughout your body so that they can be stored in your liver and fatty tissue until needed. Both dietary fat and stored fat have other important roles in the body:

Growth and Development

Fat—particularly the omega-3 fats—is needed for good health throughout the life span, beginning in the womb. A deficiency of omega-3 fatty acids during pregnancy can affect a baby's growth and development, particularly mental and visual functioning. Children require essential fats, too, to support good health during their growth years.

A Source of Energy

Fat is a fuel that provides a substantial portion of energy required to drive the body's basal metabolism, which represents the energy it takes just to exist—or, put another way, the energy needed to control vital internal functions such as breathing, heartbeat, hormone secretion, and the activity of the nervous system. When needed, stored body fat can be converted into an emergency energy supply to help us stay alive in the event of a long famine, or during a debilitating illness to provide energy to battle the disease.

Dietary fat is a highly concentrated source of energy. One gram of fat provides twice as much energy (calories) as one gram of carbohydrate or protein. So if you're hiking or hunting, particularly in cold weather, you need lots of food energy to sustain yourself—food energy that's best supplied mostly by fat-rich foods.

Fat is considered an exercise fuel, but more of a second-string source of energy. During exercise, your body prefers to burn carbohydrate for energy, as glucose in your blood or glycogen stored in your muscles. But if carbohydrate dwindles, your body draws on fatty acids for fuel. In contrast to your limited but ready-to-use glycogen stores, fat stores are practically unlimited. In fact, it has been estimated that the average adult man carries enough fat (about a gallon) to ride a bike from Chicago to Los Angeles, a distance of roughly 2,000 miles.

Flavor Enhancement

Although fairly tasteless itself, fat imparts enticing flavor and aroma to foods, thus stimulating your appetite. You experience this every time you smell bacon frying or enjoy a spoonful of ice cream. What's more, cooking with fat makes meats and baked foods more tender, moist, and flavorful. If the cookies you ate for dessert "melted in your mouth," you have fat to thank for the sensation.

Appetite Control

Although fats stimulate your appetite, they can suppress it, too. That's because fats are slow to digest and thus make you feel full after you've eaten a meal. They also stimulate the intestinal wall to secrete a satiety-control hormone called cholecystokinin (CCK), which acts on nerves in your stomach and slows the rate of digestion. CCK is also released by the hypothalamus, the body's appetite control center. The net effect of CCK's release in the body is to tell the brain you're full. Worth mentioning, too: Among fats, polyunsaturated fatty acids are more powerful than either saturated fats or monounsaturated fats in bringing on a feeling of fullness.

Fat Storage

Many dietary triglycerides (the food fat you eat) are shuttled to fat depots—in muscles, breasts, thighs, hips, the area under your skin, and other places—where they are stored. This occurs only after your body has used all the fats and carbohydrates it needs for energy.

Your body stores dietary fat much more readily than it does carbohydrates, which must first be dismantled into tiny fragments and reassembled into fatty acids—a complex process that requires lots of energy. But because dietary fat is so chemically similar to body fat, it requires fewer breakdown steps before it is stored. Therefore, if you eat the same number of excess calories from fat and carbohydrate, your body will store the fat calories rather than the carbohydrate calories. The more fat you eat, the more you're likely to wear.

It is this storage fat that we're always trying to get rid of. When you put on fat weight, storage fat cells—which are specialized for hoarding fat—become stuffed with fat and enlarge as a result. If you gain fifty pounds or more, fat cells start to multiply, and you've got them for life. Dieting doesn't obliterate fat cells either. It only shrinks them.

Protection and Insulation

But even storage fat serves some useful purposes. Some storage fat pads organs for protection and acts as a shock absorber, cushioning them from jolts. Most storage fat is found just under the skin, where it insulates us from extremes in temperature.

Storage fat makes up most of the 20 to 30 billion fat cells in our bodies. A smaller percentage of body fat—around 15 percent—is classified as "essential fat." It is the structural constituent of vital body parts such as the brain, nerve tissue, bone marrow, heart, and cell membranes. Storage fat is constructed from the essential fatty acids you obtain from your diet.

Hormone Production and Control

Fats are also required for hormone production and regulation. Produced by glands, tissues, and organs, hormones are chemical messengers that control various conditions in the body. While most hormones are secreted by glands, some are produced in fatty tissue. A good example is a lipid called 7-dehydrocholesterol, found in the fat just beneath the surface of your skin. It is activated by sunlight and converted into vitamin D, which has a number of hormonal duties in the body.

Fatty tissue is also involved in regulating the production of female sex hormones, mainly estrogen. Estrogen is the collective name for a trio of female hormones: estradiol, secreted from the ovaries during the reproductive years; estriol, produced by the placenta during pregnancy; and estrone, secreted by the ovaries and adrenal glands and found in women after menopause. These naturally occurring estrogens are responsible for developing the female sex characteristics, regulating menstrual cycles, and maintaining normal cholesterol levels.

At puberty, after a young girl has gained a certain percentage of body fat, she begins to menstruate and develop sexually. Throughout life, too much fat or too little fat can interfere with the ability to have normal periods, ovulate, and become fertile. Losing a lot of weight and depleting body stores, for example, leads to an estrogen deficiency similar to menopause, and periods cease.

A Health-Enhancing Nutrient

So you see: Fat is a valuable nutrient in your diet. And what's more, some fats are absolutely vital for good health. Without them, you're putting your well-being on the line.

As with any nutrient, the misuse of fat—whether from eating too much of it or not eating enough of certain kinds of fat—can detract from your health. In the next chapter, we'll take a look at this issue and delve into why some fats can be fatal.

• T W O •

When Fat Can Be Fatal

The reason fat has gotten such a bum rap has to do with America's number one killer—cardiovascular disease. Each day in the United States, cardiovascular disease claims more than 2,600 lives—an average of one death every thirty-three seconds, according to the most recent statistics from the American Heart Association.

The accepted explanation for such high rates of cardiovascular disease has been this: Saturated fats and cholesterol in our diets lead to high cholesterol in the blood, which in turn clogs blood vessels, contributing to heart attack and stroke. So for nearly forty years, we've been told that by shunning foods such as butter, cream, cheese, eggs, and meat—sources of saturated fats and cholesterol—and replacing them with low-fat, cholesterol-free foods, we can cut our risk of cardiovascular disease.

But there is a little bit more to the story. Emerging from decades of research is the fact that multiple factors appear to play a role in the development of cardiovascular disease: smoking, lack of exercise, high blood pressure, excess sugar consumption, poor diet quality—and fat. Thus, the fat/disease controversy is far from resolved. One thing is certain, though: The typical American diet is too high in certain types of fats with known links to disease.

In this chapter, we'll look into the fat factor in disease—and why some fats promote illness and poor health.

Saturated Fat, Cholesterol, and Cardiovascular Health

Over time, too much saturated fat in your diet can harm the health of your cardiovascular system. Essentially, excess saturated fat disrupts your liver's ability to break down excess cholesterol, a fat that is a building block for cells and hormones. Further, saturated fat causes your liver to churn out cholesterol to form an artery-clogging type of cholesterol known as low-density lipoprotein (LDL) cholesterol, dubbed the "bad" cholesterol.

For background, cholesterol comes as cholesterol in your blood and cholesterol in food. Because your body can manufacture cholesterol from fats, carbohydrates, or proteins, you don't require cholesterol from food.

When you eat a food that contains cholesterol, that cholesterol is simply broken into smaller components of various fats and proteins that are used to make many other substances that your body requires. In other words, the cholesterol you eat doesn't raise the cholesterol in your blood. It's just that some foods high in cholesterol also happen to be loaded with saturated fat. The more saturated fat you eat, the more cholesterol your liver makes.

If your liver overproduces cholesterol and pumps out LDL cholesterol, the excess circulating in the bloodstream forms lesions inside the walls of your arteries, where it is eventually deposited. LDL cholesterol is highly susceptible to oxidation, a tissue-damaging process that occurs when oxygen reacts with fat—in this case, the lipid portion of LDL cholesterol. Oxidation is thought to play a role in the formation of these arterial lesions. Your body begins forming plaque as a biological bandage to repair the lesions. Trouble starts, though, when plaque builds up in an artery, narrowing the passageway and choking off blood flow. A heart attack can occur when blood flow to the heart muscle is cut off for a long period of time, and part of the heart muscle begins to die.

A body-friendly cholesterol is high-density lipoprotein, (HDL) cholesterol. It contains the least cholesterol and does not cause lesions. Its job is to pick up the bad cholesterol from the cells in the artery walls and transport it back to the liver for reprocessing or excretion from the body as waste. HDL cholesterol has been nicknamed the "good" cholesterol.

A medical test called the *blood lipid profile* detects the amount of cholesterol in your blood, which is considered a prime forecaster of your likelihood of suffering a heart attack or stroke. But it is not the only forecaster; others are smoking and high blood pressure (hypertension).

A total cholesterol reading above 200 may be a danger sign. Generally, HDL cholesterol should be higher than 35, while LDL should be 100 or below. Borderline high LDL is considered 130 to 159; high LDL, 160; and very high LDL, 190, according to new recommendations from the National Heart, Lung, and Blood Institute. High blood LDL cholesterol is therefore a major risk factor for heart disease, but one that can be controlled by losing weight, eating less saturated fat, and exercising more.

Some people, however, have high cholesterol no matter how well they watch their diets or exercise. That's because their livers are genetically programmed to overproduce cholesterol. In such cases and in certain high-risk cases, doctors may prescribe cholesterol-lowering drugs. Most work by interfering with a liver enzyme's ability to manufacture cholesterol.

When checking cholesterol, your doctor may be on the lookout for the *lipid triad,* which also increases your risk of cardiovascular disease. Essentially, the lipid triad describes the presence of elevated triglycerides (dietary fats not fully broken down by the liver that circulate in the blood), too-low HDL cholesterol, and high LDL cholesterol—in particular, a type of LDL cholesterol characterized by its small particle size. "Small" LDL cholesterol can get into artery walls more easily and cause damage more quickly than larger LDL particles can.

Interestingly, the lipid triad worsens not by eating a high-fat diet, but by indulging in a high-carbohydrate diet, especially one that is laced with refined carbohydrates and sugary foods. Although it reduces LDL cholesterol, a high-carbohydrate diet lowers the good HDL cholesterol and elevates your triglycerides. Here's why: If you load up on carbohydrates—without burning them up—your liver will convert them into saturated fat. So eating lots of carbohydrates, even though you're forfeiting fat, doesn't do your heart much good.

Saturated Fat and Cancer

Diets high in saturated fat have been linked to certain types of cancers, largely because animal studies beginning in the 1950s began to show evidence of an association. However, this link is not an open-and-shut case.

Take breast cancer, for example—the most extensively studied cancer in terms of its relationship to dietary fat.

Until fairly recently, saturated fat was believed to be a criminal in the promotion of breast cancer. However, a study of nearly 90,000 women conducted by doctors at Boston's Brigham and Women's Hospital exonerated saturated fat, finding little evidence of the suspected breast cancer link. Other research has produced similar findings. Most investigators believe that multiple factors are at work to increase the risk of breast cancer—including genetics, menstrual history, sedentary lifestyle, body fat, and overall diet—so it's difficult to pin the cause on saturated fat alone.

Diets overloaded with saturated fats have also been implicated in the development of prostate cancer and colon cancer. With regard to prostate cancer, saturated fat is thought to alter levels of sex hormones, an environment that can promote cancer. In a study recently conducted in France, investigators found that men whose diets contained more than 30 to 40 percent fat (most of it saturated) had a higher risk of developing prostate cancer than men whose diets contained less than 30 percent fat.

The risk of colon cancer catching up with you sometime in the future may be related to the amount and type of fat you eat. A Harvard study discovered that men who ate low amounts of saturated fat (7 percent of their calories) had half the rate of precancerous polyps than men who ate double that amount (14 percent). Polyps can progress into tumors in the colon.

As for the type of fat you eat, oil from fish has been found to protect against colon cancer, in contrast to the possible cancer-promoting effect of saturated fat. What's more, in countries where people consume a lot of olive oil, rates of colon cancer are very low.

Keep in mind that the link between saturated fat and cancer is still under debate in scientific circles. Rather than point fingers at saturated fat only, scientists and nutritionists have begun to emphasize the importance of overall dietary quality, specifically the need to eat more vegetables, fruits, whole grains, fiber, and other nutrient-packed foods, in order to reduce your cancer risk.

Saturated Fat and Inflammation

Saturated fats promote the production of arachidonic acid, the fatty acid that gives rise to inflammatory agents in the body, namely bad prostaglandins (series 2) and pro-inflammatory leukotrienes. These agents can harm your joints, leading to arthritis, and can trigger abnormal blood clotting and thus promote clogged arteries. Bad prostaglandins and leukotrienes have also been implicated in migraine headaches and psoriasis. By reducing the amount of saturated fat you eat, it's quite possible to tame the production of inflammatory substances in your body.

The Ugliest of All: Hydrogenated and Trans-Fatty Acids

In truth, the real culprit in life-threatening diseases may turn out not to be saturated fat, but a super-deadly fat known as *hydrogenated* or *partially hydrogenated* fat. Examples include margarine and vegetable shortening. Hydrogenated fats are also used in commercially baked products, including cakes, doughnuts, cookies, crackers, potato chips, and other snack foods.

Hydrogenated fats are polyunsaturated omega-6 fatty acids that have been synthetically altered in a process called hydrogenation. To produce them, manufacturers take the cheapest oils available—usually soy, corn, or cottonseed—which are already rancid from the extraction process, then bubble hydrogen into the oil. Nickel oxide, which is quite toxic, is also used in the process, and traces are left in the final product. In the case of margarine, other chemicals are added (most notably bleach and coal-tar dyes) to make it look like butter.

Hydrogenation changes the chemical makeup of the fat to harden it, make it more spreadable than the original oil, and keep it fresh longer. But consequently, the unsaturated fatty acids become saturated as they take on the hydrogen and behave more like saturated fats in your body. The original fat is robbed of its unsaturated properties and healthful benefits.

There's more: During hydrogenation, some of the unsaturated fats, rather than turning into saturated fats, change their shape and morph into unnatural fats called trans-fatty acids. These fats are loathsome to your body. They slink into cell membranes, making them rigid, inflexible, and generally impaired. Consequently, the victimized cells are handicapped, unable to protect themselves against invaders or circulate freely through blood vessels.

Some major research into the effects of trans-fatty acids on health reveals the rather threatening nature of these fats. A landmark Harvard study found that women who ate four or more teaspoons of margarine a day had a 66 percent higher chance of heart disease than women who ate less than one teaspoon a month. The probable cause: Trans-fatty acids inhibit the body's ability to properly use essential fatty acids (the good fats) and elevate cholesterol.

The cholesterol-elevating problem associated with trans-fatty acids was revealed in a study conducted in the Netherlands. Investigators observed that trans-fatty acids raised levels of LDL cholesterol to the same degree that saturated fats did. And according to another study, cooking with margarine is health-risky, increasing your odds of heart disease by 90 percent. The Framingham Heart Study, a long-term study monitoring the health status of 5,127 men and women from Framingham, Massachusetts, also found that eating margarine increased the risk of heart and artery disease.

Diets overloaded with trans-fatty acids may increase your risk of breast cancer, too. That's the finding of a University of North Carolina study showing that women whose fatty tissue contained high levels of trans-fatty acids and low levels of essential fatty acids were three times more likely to develop breast cancer.

Other research has found that diets high in hydrogenated and trans-fatty acids are linked to other serious problems: prostate cancer, diabetes, obesity, immune disorders, low-birth-weight babies, lactation problems, sterility, and bone disorders. Without a doubt, you should stay away from these bad fats. Here are some guidelines for steering clear of hydrogenated and trans-fatty acids:

- Read food labels. Even foods claiming to be cholesterol-free, low-fat, or natural may contain trans-fatty acids. That being so, avoid food products that have the words *hydrogenated* or *partially hydrogenated* on their labels.

- Limit or avoid using stick margarine, which is the most highly hydrogenated fat of all. Softer tub and liquid margarines are lower in trans-fatty acids, and are a better alternative. In fact, a study conducted at the University of Texas Southwestern Medical Center revealed that tub margarine reduced LDL cholesterol by an average of 10 percent in adults and children over a five-week period.

- Butter isn't necessary a bad choice if you're a spread lover. That's because it contains no trans-fatty acids. Just use a dab, though, since butter is high in saturated fats. (See Table 4 for important information on butter.)

- Avoid cooking with stick margarine or shortening. Substitute vegetable oil. Or, for a fat-free recipe, replace the fat with applesauce or fruit purée.

- Choose margarines and other fats that contain liquid vegetable oil as the first ingredient and no more than two grams of saturated fat per tablespoon.

- Cut back on foods that are fried in vegetable shortening, such as French fries or fried chicken.

- Use olive oil in place of margarine for dipping breads, rolls, or bagels.

- Use olive oil or canola oil to sauté vegetables and other foods.

- Check out the new margarines made without trans-fatty acids. These include trans-free olive oil spreads; spreads made with a blend of soy, canola, olive, and palm oils; and trans-free, fat-free margarines and spreads (these use carbohydrate-based fat replacers rather than fat).

The Truth About Tropical Oils

Cast as fat villains on the nutritional stage are a group of saturated fats known as the tropical oils. They include coconut, palm, and palm kernel oils, and the cocoa butter in chocolate. Tropical oils are generally found in commercial baked goods and other processed foods because they impart an appealing flavor and consistency to these foods and extend their shelf life.

Coconut oil, for example, is added to some nondairy creamers and nondairy dessert toppings to replace butterfat (cream). But ironically, coconut oil is more saturated than cream. It has been shown in studies to raise cholesterol levels and is thus likely to increase the risk of heart disease. Palm kernel oil elevates cholesterol, too, and is considered risky, as well.

Palm oil, on the other hand, is much less of a villain. Though saturated, palm oil elevates the good HDL cholesterol, prevents abnormal clotting in the blood, and reduces blood pressure. Plus, it contains vitamin E, an important antioxidant that helps keep cholesterol in check. Another healthful

Table 4

THE CASE FOR BUTTER

Butter is not the bad guy it's been made out to be. Consider the following facts:

• The idea that butter elevates cholesterol has not been well substantiated by research. Margarine is a worse offender.

• Butter is a good source of fat-soluble vitamins such as vitamins A, D, and E. In fact, vitamin A from butter is more readily absorbed than from other sources.

• Chemically, butter consists of short- and medium-chain fatty acids, which are absorbed directly from the small intestine to the liver for quick energy and have less of a tendency to be stored as body fat.

• Butter contains lauric acid, the only saturated fatty acid not made by the body. Lauric acid is notable because it fights disease-causing microorganisms and tumors, and it helps bolster the immune system. In excess, however, lauric acid increases cholesterol levels in the blood.

• Butter contains two very-short-chain fatty acids—proprionic acid and butyric acid—that have antifungal and antitumor powers.

• Butter is rich in selenium, an antioxidant mineral.

• A natural constituent of butter is lecithin, a phospholipid involved in the proper metabolism of cholesterol.

component of palm oil is oleic acid, a monounsaturated fatty acid plentiful in olives and olive oil. Oleic acid has beneficial effects on cholesterol levels.

As for cocoa butter—the fat found naturally in chocolate—the news is good. Studies show that it does not adversely affect heart health the way many other saturated fats do. There's more: A Harvard study found that men who ate chocolate one to three times a month lived longer than men who ate none.

Does this mean chocolate is a health food?

Not exactly. However, cocoa butter is rich in a saturated fatty acid called stearic acid, also found in meat and dairy products. Research shows that stearic acid does not elevate cholesterol, nor does it promote heart attacks. Further, it prevents the formation of dangerous blood clots.

This encouraging information isn't a license to go overboard on chocolate, however. Scientists and nutritionists advise that chocolate can still be savored, but in moderation, since it is high in calories and fat.

When it comes to tropical oils, there are villain fats (coconut oil and palm kernel oils) and hero fats (palm oil and cocoa butter). Many food manufacturers, however, have removed tropical oils from cookies, cakes, and crackers. A word of caution, though: Some of these fats, particularly coconut and palm oils, will show up as hydrogenated or partially hydrogenated fats in various food products. Read labels, and try to avoid these types of fats.

When Polyunsaturated Fats Go Bad

Generally, when saturated fats are replaced in the diet with polyunsaturated fats—specifically those of the omega-3 and omega-6 families—cholesterol levels drop. But the problem is that polyunsaturated fats, particularly the omega-6 fatty acids, cut all cholesterol—the bad kind (LDL) and the good kind (HDL). Omega-6 fats are found in commercial vegetable oils made from corn, soy, safflower, sunflower seeds, and peanuts. To make matters worse, cancer researchers have discovered a close correlation between diets high in omega-6 fatty acids and a greater risk of cancer.

But aren't polyunsaturated fats supposed to be good for you?

Yes—they are among the healthier fats. But a problem with polyunsaturated fats is that they are highly vulnerable to oxidation, a process in which fat is exposed to oxygen and turns rancid as a result. As polyunsaturated fats sop up oxygen, they can be attacked by free radicals and converted into harmful molecules called *lipid peroxides*. Peroxides attack cell membranes, setting off a chain reaction that creates many more free radicals. Pits form in cell membranes, allowing harmful bacteria, viruses, and other disease-causing agents to gain entry into cells. Other structures such as body proteins, DNA (the genetic material inside cells), and cartilage can be attacked by free radicals and damaged too. The result is a frenzy of cell destruction. Lipid peroxidation is believed to be one of the mechanisms involved in the development of autoimmune diseases, heart disease, and cancer.

Food processing encourages oxidation. Thus, foods most likely to contain rancid, oxidized polyunsaturated fats are processed food products—crackers, snack foods, frozen foods, cookies, pastries, baked goods, and packaged foods, among others.

Always read the labels of processed foods. Shun them if the label lists polyunsaturated fats such as safflower oil, sunflower seed oil, corn oil, soybean oil, or peanut oil, because you can bet they are oxidized. Opt for fat-free foods instead.

Another problem, specifically with omega-6 fats, is that in excess they stimulate the formation of arachidonic acid, which is a precursor or building block of "bad" prostaglandins and pro-inflammatory leukotrienes that are involved in inflammation. Although we need some arachidonic acid, too much is believed to be responsible for the rise in arthritis and other chronic inflammatory diseases. An oversupply of arachidonic acid also fuels the growth of cancer cells.

Risky Business: Low-Fat Diets

If you're a Jack Sprat—that is, you eat little or no fat—you risk an essential fat deficiency, which can be as health-damaging as eating too much of the wrong kind of fat. Case in point: A study published in the medical journal *Metabolism* found that in forty-seven patients with heart disease, blood levels of essential fatty acids were significantly lower than levels found in healthy people. Ouch!

Further, a study in the *British Journal of Nutrition* showed that a low-fat diet can affect your mood—in a bad way. In this study, ten men and ten women ate a higher-fat diet (41 percent of calories from fat) for one month, then switched to a low-fat diet (25 percent of calories from fat). While on the higher-fat diet, volunteers experienced less tension and anxiety, suggesting that cutting fat may make you feel down in the dumps. The study didn't specify why this may have occurred, but it is well known that essential fats promote good mental functioning.

Some scientists feel that essential fatty acid deficiency may be the most serious dietary health problem facing Americans. With fat-slashing diets, the body has trouble absorbing the fat-soluble vitamins A, D, E, and K. Furthermore, the health of cell membranes is jeopardized because low-fat diets are low in vitamin E. Vitamin E is an antioxidant that prevents disease-causing free radicals from puncturing cell membranes.

So valuable are essential fatty acids that deficiencies can cause a wide range of serious symptoms and illnesses. Table 5 lists the physical conditions that can be brought on by a short supply of essential fatty acids. Clearly,

Table 5

ESSENTIAL FATTY ACID DEFICIENCY: SYMPTOMS AND ILLNESSES

Fatigue	Arthritis
Dry skin and hair	Chest pain
Cracked nails	Cardiovascular disease
Dry mucous membranes (mouth, tear ducts, vagina)	High blood pressure
	Memory problems
Digestive problems	Depression
Constipation	
Poor immunity	
Frequent colds	
Joint problems	

you'll want to make sure your diet provides ample essential fats for good health.

Putting Bad Fats in Perspective: A Disease-Prevention Strategy

Saturated fats and other bad fats should not receive all the blame for heart disease and other life-shortening illnesses. Few pieces of evidence bring this more clearly into focus than a landmark study published in the *New England Journal of Medicine* in 2000. For fourteen years, Harvard researchers followed more than 84,000 women participating in the Nurses' Health Study. They were free of heart disease, cancer, and diabetes in 1980—when the study began. By the end of the study, women at the lowest risk of disease, particularly heart disease, were those who did not smoke and exercised at least a half hour daily. Further, they followed diets that were high in fiber, folic acid (a B-vitamin that protects against heart disease), and good fats (those from fish and vegetables), and were low in trans-fatty acids and sugar.

Translation: It's the totality of your lifestyle, not just one unhealthy portion of it, that most affects your disease risk. That being the case, here's what you can do to stay as healthy as you can, for as long as you can:

Reduce your intake of saturated fats and trans-fatty acids in favor of unsaturated fats

According to the American Heart Association, the maximum amount of fat considered healthy in your daily diet is 30 percent or less, based on the number of calories you eat over several days, such as a week. To translate this recommendation into meaningful terms, a 2,000-calorie diet should contain about sixty-five grams of fat a day.

Saturated fat and trans-fatty acids combined should be 7 to 10 percent or less of total daily calories; polyunsaturated fats should also be at 10 percent or less; and monounsaturated fats should make up to 15 percent of total calories. Dietary cholesterol should be kept to a daily maximum of 300 milligrams or less (200 milligrams or less if you suffer from high cholesterol).

Quit smoking

Each year, smoking kills approximately 500,000 Americans. Heart attack, lung cancer, and chronic lung disease are the chief smoking-related diseases that claim human life. Yet death from smoking doesn't have to happen. It is the single largest preventable cause of premature death and disability.

Maintain a normal weight for your height and body frame

After smoking, weight-related conditions are the second leading cause of death in the United States, claiming 300,000 lives each year. Although overweight and obesity are considered to be appearance problems, they are in fact serious conditions, directly linked to a number of disabling and life-threatening diseases. Among them: coronary heart disease, stroke, some cancers, diabetes, high blood pressure, gallbladder disease, osteoarthritis, and mental health problems. The best way to determine a normal weight is to have your body fat percentage tested. Generally, a body fat percentage between 18 and 24 percent for women and between 15 and 18 percent for men is considered optimal.

Exercise at a moderate-to-vigorous pace for at least thirty minutes most days of the week

An estimated 250,000 deaths in the United States each year are linked to a lack of exercise, according to the Centers for Disease Control and

Prevention. It's well established by research that exercise protects you against heart disease, cancer, obesity, bone diseases, and many other life-limiting illnesses.

Fill up with fiber

Most plant foods, including cereals, pasta, fruits, and vegetables, are complex carbohydrates that are packed with dietary fiber, an indigestible carbohydrate that has a long list of impressive health benefits. Fiber has a cholesterol-lowering effect, and it helps relieve constipation, rid the body of cancer-causing substances, and assist in weight control. The National Research Council recommends 20 to 35 grams of fiber a day.

Eat fish two to three times a week

Fish contains beneficial fats that are endowed with numerous health benefits. You'll learn more about the health-boosting powers of fish and fish oils in chapter 3.

Obtain enough folic acid

Found in green leafy vegetables, this B-vitamin reduces homocysteine, a proteinlike substance, in the tissues and blood. High homocysteine levels have been linked to heart disease. Scientists predict that as many as 50,000 premature deaths a year from heart disease could be prevented by increasing consumption of folic acid.

Recent scientific experiments have revealed that folic acid deficiencies cause DNA damage that resembles the DNA damage in cancer cells. This finding has led scientists to suggest a link between cancer and a folic acid deficiency. Other studies show that low levels of folic acid appear to be associated with premalignant cell growth in the cervix. The recommended daily intakes of folic acid are as follows: women, 400 micrograms; pregnant women, 600 micrograms; lactating women, 500 micrograms; and men, 400 micrograms.

Reduce your intake of refined carbohydrates

These foods include sugar, sweets, breads, and processed snack foods. Such foods provoke metabolic problems that promote fat storage, elevate blood fats, and raise blood pressure. Sugar, in particular, lowers your body's resistance to disease and creates deficiencies in heart-protective B-vitamins.

Part II

Omega Healing

· THREE ·

Fishing for Good Health

Question: What do heart disease, arthritis, cancer, diabetes, bowel disease, psoriasis, and depression have in common?

Answer: All of these troubling ills can be treated effectively with a group of essential polyunsaturated fats called omega-3 fatty acids. Omega-3 fatty acids come in three varieties: alpha-linolenic acid (ALA), eicosapentaenoic acid (EPA), and docosahexaenoic acid (DHA).

Alpha-linolenic acid, which must be obtained from your diet, is found mostly in plant foods such as flax, soybeans, and vegetables. From these foods, it is converted to EPA and DHA in your body. EPA and DHA are also found in fish, where they are *preformed*. This simply means that they don't require conversion from alpha-linolenic acid and thus are better utilized by the body. You can thus obtain EPA and DHA directly from fish and shellfish.

Seafood is the richest source of EPA and DHA in our food supply. Cold-water fish such as salmon, mackerel, and tuna supply the most EPA and DHA because they swim around in chilly waters and develop a thick layer of insulating fat to keep warm. If not for the abundant unsaturated fatty acids in these fish, their fat would solidify in the cold-water environment, and the fish would be unable to swim.

In addition, cold-water fish feed on plankton, which contains alpha-linolenic acid (the original aquatic source of EPA and DHA) and algae, which is chock-full of DHA. But if you're not a fish eater, don't worry. You can obtain EPA and DHA from dietary supplements. Table 6 provides information on the amount of omega-3 fatty acids found in various food sources and supplements.

Healing Power from Fish

For more than twenty-five years, research reports have been pouring in over the potent health benefits of omega-3 fatty acids. Their healthful properties first came to light when scientists discovered that Greenland Eskimos have a lower rate of heart disease and stroke than other populations do, despite their high-fat diet. The difference is that Eskimos eat twenty times more fish than Americans, and fish is loaded with omega-3 fatty acids.

Because of this link, omega-3 fatty acids have become best known for their good deeds in cardiovascular health, where they have been found to lower blood pressure, reduce cholesterol, thwart dangerous blood clotting, and protect against irregular heartbeats.

Basically, your body uses omega-3 fats to manufacture eicosanoids (prostaglandins and leukotrienes), hormonelike substances that regulate many chemical processes. As noted in chapter 1, prostaglandins come in two varieties: the good prostaglandins, which have healing properties, and the bad prostaglandins, which promote disease and inflammation. Likewise, there are less inflammatory leukotrienes and pro-inflammatory leukotrienes. An imbalance of fat in the diet—namely, too much saturated fat and omega-6 fatty acids—promotes the production of bad prostaglandins and leukotrienes.

By contrast, increasing omega-3 fatty acids through diet steps up the production of good prostaglandins and less inflammatory leukotrienes in the body. As a result, omega-3 fats can alter your metabolism in four positive ways: They slow the rate at which your liver manufactures triglycerides; they make your blood less sticky, so that clots are less likely to form; they help repair tissues that have been damaged by lack of oxygen (a condition that occurs when arteries become blocked and can't deliver oxygen to the heart or brain); and they help lower blood pressure, a risk factor for heart attacks and stroke. In addition, omega-3 fats act like white knights of sorts, protecting the body when its own immune system attacks tissues, as in rheumatoid arthritis.

Table 6

OMEGA-3 FATTY ACID CONTENT OF FOODS

	Food (per 100 grams)	ALA (grams)	EPA + DHA (grams)
Oils	Perilla oil	63.6	0
	Flaxseed oil	53.3	0
	Black currant oil	12–14	0
	Canola oil	11.1	0
	Soybean oil	6.8	0
Nuts and Seeds	Walnuts, English	6.8	0
	Walnuts, black	3.3	0
	Soybean kernels	1.5	0
Vegetables	Soybeans, green, raw	3.2	0
	Soybean sprouts	2.1	0
	Purslane	0.4	0
	Beans (navy, pinto)	0.3	0
Fish	Sardines, in sardine oil	0.5	3.3
	Atlantic mackerel	0.3	2.5
	Atlantic salmon	0.1	1.8
	Pacific herring	0.1	1.7
	Atlantic herring	0.1	1.6
	Lake trout	0.4	1.6
	Bluefin tuna	trace	1.6
	Anchovy	trace	1.4
	Atlantic bluefish	trace	1.2
	Pink salmon	trace	1.0
	Bass, striped	trace	0.8
	Florida pompano	trace	0.6
	Halibut, Pacific	trace	0.4
	Catfish	trace	0.3
	Cod	trace	0.3
	Flounder	trace	0.2
	Haddock	trace	0.2
	Red snapper	trace	0.2
	Swordfish	trace	0.2

Table 6 (continued)

OMEGA-3 FATTY ACID CONTENT OF FOODS

	Food (per 100 grams)	ALA (grams)	EPA + DHA (grams)
Shellfish	Alaska king crab	trace	0.3
	Shrimp	trace	0.3
	Lobster	trace	0.2
Mollusks	Oyster	trace	0.6
	Mussel, blue	trace	0.5
	Scallop	trace	0.2
	Clam	trace	trace
Supple-ments	Promega		44.2
	MaxEPA		29.4
	Salmon oil		19.9
	Cod liver oil		18.5

Adapted from: Nettle, J.A. 1991. Omega-3 fatty acids: comparison of plant and seafood sources. *Journal of the American Dietetic Association 91*: 331–337; Connor, W.E., et al. 1993. (N-3 fatty acids from fish. *Annals of the New York Academy of Sciences 14*: 16–34.)

Thus, omega-3 fats have been heralded for their ability to treat an amazing array of other diseases, from heart disease to cancer. Here's a closer look.

Battle Heart Blockages

Elevated levels of cholesterol and triglycerides are risk factors for heart disease. In excess amounts, cholesterol can collect in the inner lining of the arteries and lead to a heart attack. Triglycerides are a type of fat that circulates in your blood. When levels are too high, HDL cholesterol (the good kind) tends to fall. Omega-3 fats have been found to influence both cholesterol and triglycerides. A few examples:

In one study, patients who supplemented with three tablespoons of fish oil every day reduced their total cholesterol by 15 percent in just four weeks. One specific type of fish oil—salmon oil—reduces triglycerides, cholesterol, and a harmful type of cholesterol known as VLDL (very-low-density lipoprotein) in people with an excess of blood fats in their system.

Salmon oil has also been found to increase HDL cholesterol, the beneficial kind.

But in other cases, fish oil has performed poorly as a cholesterol-lowering agent. A Mayo Clinic statistical study, called a meta-analysis, of about twenty-one trials involving people with type II diabetes found that while fish oil (ranging from 3 to 18 grams daily) slashed triglyceride levels, it unfortunately raised levels of harmful LDL cholesterol, particularly in individuals who used the higher doses. Other studies have found that fish oil has very little effect on cholesterol except when patients have elevated triglycerides.

On a more positive note, some experiments have discovered that fish oil can abort or prevent the formation of lesions in the walls of your arteries. Caused by oxidized LDL cholesterol, these lesions make it possible for cholesterol to collect in the arteries and eventually plug them up. Fish oil thus protects your arteries from damage and saves them from potential blockages.

It's vital to note that eating fish rich in omega-3 fats, rather than taking fish oil supplements, may be a better prescription for keeping your arteries healthy. Canadian researchers found that people who ate more than eight ounces of fish a week stood a better chance of having their arteries stay open after angioplasty (a procedure that dilates arteries) than patients who ate no fish. Another study showed that taking fish oil capsules did not keep arteries open following angioplasty.

Regulate Normal Blood Clotting

There are two chemicals in your body that, in the right balance, are responsible for preventing abnormal blood clots but assuring the prompt formation of clots when you're cut or bruised, in order to control bleeding and help you heal. These chemicals are thromboxane and prostacyclin.

Thromboxane, a prostaglandin mainly synthesized from arachidonic acid, is the chemical that tells platelets—tiny clotting factors in your blood—to clump together, forming blood clots. But should one of these clots wend its way into a blood vessel narrowed by atherosclerosis (hardening and thickening of the arteries), it could trigger a heart attack or stroke. Prostacyclin, synthesized mostly from EPA, orders platelets to not stick together and move along.

If a platelet has too much arachidonic acid in its cell walls, more thromboxane is produced, and the platelet will clot too easily, possibly leading to heart attack or stroke. By contrast, if a platelet has enough EPA in its cell wall, less thromboxane is made, prostacyclin becomes more potent, and blood stays more fluid.

Supplying your body with ample amounts of omega-3 fats keeps thromboxane and prostacyclin production in proper balance. Consequently, dangerous clots are less likely to form.

Heal Hypertension

Known as the silent killer, high blood pressure (hypertension) afflicts about 60 million Americans, half of whom do not know they have it. Hypertension is a serious condition because it contributes to heart attacks and strokes. Blood pressure exceeding 140/90 spells danger and should be brought under control. Prescriptions for high blood pressure include exercising, restricting salt intake, cutting down on alcohol and caffeine, quitting smoking, losing body fat, and taking blood pressure medication.

To that list, we can add consuming fish and fish oil—and justifiably so. Taking 3,000 milligrams of fish oil (the amount in seven ounces of fatty fish) a day reduces high blood pressure, according to a meta-analysis of seventeen controlled clinical trials of the effect of fish oil supplements on high blood pressure.

In another study, volunteers ate a mackerel diet containing about 5 grams of omega-3 fats every day. Their blood pressure fell moderately, from 152/93 to 140/89, on average. With slightly more omega-3s (6 grams daily), blood pressure may drop even lower, from 147/82 to 124/74, according to other research.

It's not clear exactly how fish and fish oil reduce blood pressure, but scientists theorize that the same chemicals—thromboxane and prostacyclin—that help control blood clotting are believed to lower blood pressure. Thromboxane constricts blood vessels, while prostacyclin dilates them. When in balance, these chemicals help arteries relax and stay flexible, causing a reduction in blood pressure.

Protect Against Irregular Heartbeats

Each year in the United States, about 300,000 people die within one hour of suffering a heart attack, primarily due to irregular heartbeats (arrhythmias) stemming from damage to heart tissue. Arrhythmias occur when the electrical system that controls your heartbeat is disturbed. The ventricles, the lower chambers of your heart, begin to contract rapidly and chaotically, resulting in insufficient blood flow to your vital organs.

Evidence is mounting that a simple dietary change—eating more omega-3 fats—may help prevent fatal arrhythmias. Research with isolated cardiac cells of rats shows that omega-3 fats enhance the electrical stability of heart cells. Specifically, omega-3s help regulate the orderly flow of calcium, sodium, and other charged particles into heart cells to ensure normal contractions.

In a study involving human volunteers, people took 4.3 grams of a fish oil supplement every day for sixteen weeks. At the end of the study, the fish oil–takers had a marked reduction in the average number of arrhythmias— from 5.9 to 2.9.

Eating fish also helps keep the heart from beating irregularly. Case in point: 295 men who ate just one fatty-fish meal a week had half the risk of heart failure due to arrhythmia.

No one can yet guarantee that fish or fish oil will fully prevent arrhythmias, but if you want to hedge your bets, eat more fish or consider supplementing with fish oil.

Relieve Arthritis

One of the most promising uses of fish and fish oil has been in the treatment of rheumatoid arthritis, a joint disease in which the body's own immune system attacks its tissues. More than 2 million people have this form of arthritis. Most are women.

The chief symptoms of rheumatoid arthritis are pain, swelling, and stiffness—all thought to be triggered by pro-inflammatory prostaglandins called leukotrienes, which are synthesized from arachidonic acid. In a rather dramatic nutritional rescue, omega-3 fatty acids interfere with the conversion of arachidonic acid to leukotrienes, thus reducing their levels in the body and rendering them less inflammatory.

In a study conducted at Albany Medical College, sixty-six rheumatoid arthritis patients were given either fish oil supplements (130 milligrams per kilogram of body weight) or nine capsules of corn oil, every day for thirty weeks. All patients continued to take their regular prescription medicine during the study. As the study progressed, joint tenderness and morning stiffness decreased significantly in patients taking fish oil supplements. Some of the patients were able to stop taking their prescription medicine due to the pain-relieving effects of fish oil supplements.

This study is just one of the many trials that have been conducted, showing such encouraging results. Generally, fish oil provides moderate relief from symptoms and makes patients less reliant on anti-inflammatory drugs and pain medication. The effective dose for arthritis relief appears to be 3 to 5 grams of omega-3 fatty acids a day. That's the equivalent of eating about 8 ounces of omega-3-rich fish every day.

If you want to cut your chances of developing rheumatoid arthritis, you can get by with less—1.6 grams of omega-3 fatty acids per day or two or more servings of fish per week, according to recent studies.

Here's more good news for your joints: Omega-3 fatty acids may protect you against a more prevalent form of arthritis called osteoarthritis (a joint disease in which cartilage gradually deteriorates). Again, the proof comes from Eskimos, who have the lowest rates of osteoarthritis in the world. As noted earlier, they eat a diet rich in fish oils, namely omega-3 fatty acids, which appear to have a protective effect on joints. But in Eskimos who become "westernized," the rate of osteoarthritis triples.

Another beneficial source of joint-soothing omega-3 fatty acids is the green-lipped mussel, a dietary staple of people living in coastal areas of New Zealand. Not coincidentally, these New Zealanders have very low rates of arthritis compared to their inland neighbors, and scientists have long believed that the mussel is the reason. For nearly thirty years, extracts made from the green-lipped mussel have been a popular curative for arthritis in New Zealand and Australia.

One of these extracts is Lyprinol, a supplement derived from the green-lipped mussel that has been fairly well researched as an anti-inflammatory treatment for various diseases, including arthritis. Studies indicate that Lyprinol (210 milligrams daily) relieves symptoms associated with rheumatoid arthritis and osteoarthritis, with few side effects. It appears to work by interfering with the synthesis of a troublesome inflammatory leukotriene called B4, making it less inflammatory.

Fight Cancer

The cancer-fighting potential of omega-3 fatty acids has been studied for nearly twenty years, mostly in lab animals. Still, the findings have been exciting, particularly in the ability of omega-3s to suppress tumor growth.

A large body of animal studies shows that omega-3 fats slow the growth and spread of tumors, prolong survival, and even prevent tumors from forming.

Fish and fish oil may powerfully thwart breast cancer, too, according to population studies of large groups of people living in various regions of the world. Scientists have found that rates of breast cancer are lowest in Japan and other countries where women eat the most fish.

Scientists theorize that omega-3 fatty acids work by countering the harmful effects of bad prostaglandins, which lower immunity and promote cancer growth. When levels of these not-so-friendly prostaglandins are too high, the body can't protect itself and tumors tend to grow faster.

As research into the antitumor activity of omega-3s continues, it will be fascinating to see how it all shakes out. But until more information is available, it does seem a wise idea to eat more seafood on a regular basis.

Counter the Complications of Diabetes

The value of fish and fish oil in treating diabetes, a blood-sugar-metabolism disorder, is in countering the life-threatening complications of the disease. One of these is heart disease, the leading cause of death in people who suffer from diabetes. In fact, the death toll among diabetics with heart disease is about two to four times as high as those adults without diabetes. One reason is that people with diabetes often have high levels of blood lipids (cholesterol and triglycerides) and thus are at greater risk of developing atherosclerosis.

Fish oil has the most profound effect on lowering triglycerides—often dropping by as much as 30 percent. Conducted in the Netherlands, a statistical study of twenty-six clinical trials involving diabetics discovered that fish oil doses (ranging from 1.8 to 20 grams daily) slashed triglyceride levels, but generally had little effect on LDL or HDL cholesterol. The researchers suggested that if you have diabetes, there's room for fish oil supplements in treating triglycerides that are on the high side, particularly if standard therapies fail.

Ease Bowel Disease

In addition to heart disease and osteoarthritis, there's something else Eskimos rarely get: inflammatory bowel disease (specifically Crohn's disease and ulcerative colitis). Crohn's disease is painful, chronic inflammation of the intestine, and though rare, it seems to be on the rise. Ulcerative colitis is also a chronic condition, characterized by tiny ulcers in the inner lining of the colon.

Quite a few studies show that omega-3 fatty acids provide relief from both conditions. They work primarily by reducing the levels of that rather nasty leukotriene called B4, which is elevated in the gastrointestinal tracts of people with inflammatory bowel disease. If you have either of these diseases, try eating more fish or taking fish oil supplements, while adhering to your regular medical treatment.

Treat Psoriasis

Affecting approximately 3 million Americans, psoriasis is a genetic disease in which the life cycle of skin cells fast-forwards abnormally, and the result is skin eruptions and scaling. Psoriasis typically affects the elbows, knees, trunk, and scalp. Conventional treatment involves medications and phototherapy (exposure of the skin to ultraviolet light). But research suggests that fish oil, particularly eicosapentaenoic acid (EPA), is also an effective way to manage the disease.

People with psoriasis have higher-than-normal amounts of inflammatory leukotrienes, namely B4, in their bodies. These are believed to trigger the characteristic inflammation and scaly skin of this disease. In one study, twenty-eight psoriasis patients each were given 1.8 grams of EPA, or an olive oil placebo, every day for two months. By the end of the experimental period, itching and redness were reduced in people taking the EPA supplement. Other research indicates that EPA supplements taken with etretinate (a drug that treats psoriasis) work better than the drug by itself. If you're under the care of a doctor for psoriasis, it might be worth adding fish oil supplements to your treatment to get some relief from the itching and inflammation.

Defeat Depression

Depression, a mental illness that affects one of every four Americans, is a very common mental disorder. In fact, it has been dubbed "the common cold" of psychiatric problems—but it is the most treatable, provided the sufferer seeks treatment.

One form of treatment now getting a lot of attention in scientific circles is supplementation with omega-3 fatty acids. By studying the differences between people who get depressed and those who don't—a type of research known as psychiatric epidemiology—scientists first discovered the fascinating link between omega-3 fats and mood. Epidemiological studies have found that countries with the highest rate of fish consumption—most notably New Zealand and Japan—have very low rates of depression. Because of this knowledge, doctors and mental health experts are now suggesting that we eat more fish for protection against depression. One type of omega-3 fat—DHA—is particularly effective for escaping depression. You'll learn more about this important fatty acid in the next chapter.

A landmark study conducted at Harvard suggested that fish oil supplements (6 grams daily) can relieve symptoms of manic depression, also known as bipolar disorder, a form of depression characterized by unpredictable mood swings.

Scientists also believe that one of the instigators of postpartum depression—the down-in-the-dumps feeling experienced by some mothers a few days after delivery—is the depletion of maternal omega-3 fatty acids, a common nutritional deficiency during pregnancy. In a study of fish consumption in twenty-seven countries and the development of postpartum depression, researchers found that the more fish women ate, the less they experienced this form of depression.

Exactly how omega-3 fats help alleviate mood disorders such as depression and bipolar disorder is a puzzle. But there are some clues. Some studies suggest that higher levels of essential fatty acids in plasma may lead to increased levels of neurotransmitters—brain chemicals that transmit messages from one nerve cell to another—particularly serotonin. Serotonin is known as the "happiness neurotransmitter" because elevated levels bring on feelings of tranquillity, calm, and emotional well-being. Low levels may increase the risk for depression.

Another theory holds that the inflammatory response runs amok in people with mood disorders, and as a result, the body starts cranking out too many disease-fighting substances. When in excess, these substances hurt normal cell function and reduce levels of serotonin and other neurotransmitters. But because of their ability to thwart inflammation, omega-3 fatty acids may put the brakes on out-of-control inflammatory processes.

Other Benefits

In addition to protection against disease and depression, omega-3 fatty acids may boost your performance if you're an exerciser. In a study of thirty-two healthy men, researchers found that supplementation with omega-3 fatty acids boosted aerobic power almost as much as aerobic exercise itself.

Also, omega-3 fatty acids have the ability to dilate the capillaries. This improves the flow of oxygen and nutrients to muscles during exercise and facilitates the removal of waste products. The more nutrients the muscles can get, the better the conditions are for growth and repair.

Because of their role in the production of good prostaglandins, omega-3 fatty acids may also reduce exercise-caused inflammation and allow your muscles to repair faster following exercise.

The Promise of Shark Liver Oil

For more than forty years, one specific type of fish oil—shark liver oil—has been used medicinally as an alternative treatment for various diseases, including cancer. Its most familiar use, however, is as an ingredient in topical hemorrhoid creams. Shark liver oil is also available in supplements sold in health food stores.

A longtime popular therapeutic agent in Europe, shark liver oil is a major natural source of two beneficial compounds—squalene (see chapter 7 on olive oil) and alkylglycerols. Alkylglycerols are a group of lipids similar in structure to triglycerides. Found in fatty fish, as well as in human bone marrow and breast milk, alkylglycerols have been scientifically studied since the 1930s for their ability to reduce radiation damage, suppress tumor growth, build blood, and accelerate wound healing.

More recently, they have been researched for their role in stimulating the immune system by empowering a type of immune cell called a *macrophage*. Macrophages are "search and destroy" cells on the lookout for foreign invaders, particularly bacteria and tumor cells. Before macrophages go out on their search-and-destroy mission, they must be activated by other immune cells. It's a rather complicated process, but once activated, macrophages can secrete some sixty substances designed to kill disease-causing agents. Scientists have discovered that alkylglycerols are capable of activating macrophages, too. This discovery has prompted research into alkylglycerols as antitumor compounds.

In various studies, alkylglycerols do appear to be useful partners in cancer treatment. For example, when breast cancer cells were treated in lab dishes with chemotherapy or with chemotherapy and alkylglycerols, the combination treatment reduced tumor cells in eight out of nine patient samples. In another study, alkylglycerols inhibited tumor growth and tissue damage after radiation in women undergoing treatment for cervical cancer. Experiments have also found that patients with uterine cancer who were treated with alkylglycerols prior to radiation treatment lived longer and had a higher rate of tumor regression.

The effectiveness of alkylglycerols in reducing radiation damage may have to do with their antioxidant power, scientists believe. Although a useful therapy, radiation sets in motion a process in which the body begins spewing out devilish free radicals called *hydroxy radicals*, capable of attacking whatever they contact. Hydroxy radicals have been implicated in numerous diseases, as well as in aging. Research suggests that alkylglycerols inhibit the formation of these cellular renegades, or neutralize them at the moment of their genesis.

Numerous supplement companies manufacture shark liver oil capsules. The usual dosages are capsules with 250 to 500 milligrams of shark liver oil containing 20 percent alkylglycerols (50 to 100 milligrams of alkylglycerols). For supportive therapy during conventional cancer treatment or for immune-deficient diseases, medical experts familiar with shark liver oil recommend taking three to six 500-milligram capsules daily; for immune support and disease prevention, two to three 250-milligram capsules daily. Be sure to get your physician's blessing before supplementing, however.

Omega-3 Healing Power from Plants

As noted earlier in this chapter, alpha-linolenic acid is an omega-3 fat found mostly in plants and seed oils. Alpha-linolenic acid is a *precursor*, or building block, of EPA and DHA. However, nutritionists will tell you that it is better to get EPA and DHA directly from fish because they are better absorbed and utilized by your body.

Even so, there are health benefits to eating vegetables high in alpha-linolenic acid, since it affords significant protection against heart disease, stroke, and other illnesses. Foods rich in alpha-linolenic acid include flaxseed and perilla oil. You'll learn more about these important foods in chapter 5.

The All-Important Omega Ratio

Historically, our ancestors ate higher amounts of omega-3 fatty acids than we do today. In fact, our intake of omega-3 fats is pitifully low—which is why rates of heart disease, cancer, and other dreaded diseases may be escalating. The blame for the shift in omega-3 fat consumption can be placed mostly on food processing, which has dumped dangerous trans-fats into our foods through the process of hydrogenation. We're also eating too many omega-6 fatty acids from vegetable oils. That's bad because excess omega-6 fats interfere with the conversion of alpha-linolenic acid to EPA and DHA, and they spur the production of bad prostaglandins and leukotrienes. Our overindulgence in saturated fats has also displaced omega-3 fats in our diets.

Currently, we eat about 12 grams of omega-6 fats a day to 1.5 grams of omega-3 fats. This fat imbalance has led many nutrition and medical experts to recommend increasing omega-3 fats in our diets to achieve a more balanced ratio between omega-3 fats and omega-6 fats. Some authorities have suggested a 1:1 ratio, which would put as much omega-3 fats in our cells as omega-6 fats. This is the ratio found in the traditional diet of Greece, where deaths due to heart disease and cancer are among the lowest in world. Even by cutting our omega-6 intake in half and doubling our omega-3 intake, we'd achieve a more healthy balance.

How can we restore the balance?

There's really no magic dietary formula. Simply change your present diet to include more seafood and vegetables, while cutting back on processed foods and saturated fats, and you'll automatically balance the ratio in favor of omega-3 fats in your diet.

Eating More Fish

Generally, the best way to harness the healing power of omega-3 fats is to eat more seafood. Here are some guidelines:

- Eat two to three three-ounce fish meals a week to decrease your odds of heart disease, cancer, arthritis, and other illnesses. Just a four-ounce portion of salmon twice a week, for example, serves up about 5 grams of omega-3 fatty acids, the amount recommended by most health care practitioners.

- Not all types of fish supply the same amount of omega-3 fats. It's best to choose fattier fish such as salmon, tuna, mackerel, or sardines most of the time, since they contain the most omega-3 fats.

- Substitute fish for red meat. Research shows that this practice can alter your cholesterol profiles for the better.

- Prepare fish by broiling, grilling, microwaving, baking, or poaching. Frying can be self-defeating because the fish is often cooked in not-so-healthy oils.

- Shellfish is not off-limits if you have high cholesterol. Research shows that shellfish, which contains cholesterol, has a minimal effect on cholesterol levels. So for variety, don't be afraid to enjoy crab, oysters, shrimp, scallops, and other shellfish.

Some Fishy Precautions

Some species of fish may be contaminated with toxins, so if you eat too much of them to get your omega-3 fats, you might be ingesting too many chemicals and pollutants. Some examples of fish that may be high in toxins include grouper, marlin, orange roughy, swordfish, and shark. According to the FDA, these species have tested for high levels of mercury (considered a poison) and should be eaten no more than twice a week, or, if you're pregnant, no more than once a month.

To be on the safe side:

- When purchasing fish, select a smaller fish within the species. It is typically younger and probably hasn't been exposed to toxins for as long as older, larger fish have.

- Buy farm-raised fish, if available. These are raised under controlled conditions, with less exposure to toxins and bacteria.

- Don't eat the skin or fatty portion of fish, because this is where toxins tend to congregate.

- Avoid eating the same species of fish all the time, to minimize possible exposure to the pollutants over and over again. In other words, plan your diet to include a variety of fish.

Using Fish Oil Supplements

Most doctors and medical researchers suggest that you eat more seafood—at least two to three fish meals a week—rather than rely on fish oil supplements. But if you don't like fish, supplements may be a good option. Some words to the wise if you choose this alternative:

- Supplementing with 3 grams of fish oil daily from food sources and/or dietary supplements is considered a safe dose by the Food and Drug Administration (FDA). But taking more than 3 grams daily may thin your blood and prevent it from clotting normally. Higher doses should be taken only under medical supervision.

- Fish oil supplements are not without other side effects. An excess of these oils can be harmful and cause internal or external bleeding. Other side effects include belching, gas, nausea, heartburn, diarrhea, and fish odor on your breath or from your body.

- Take fish oil supplements with your meals to help minimize fishy breath and heartburn.

- Being a fat, fish oil supplements are high in calories and dietary cholesterol. A single capsule or softgel, for example, contains roughly 15 calories; some recommended dosages (multiple capsules daily) supply 200 extra calories a day. Unless you figure this into your total daily caloric intake, you could pack on unwanted pounds. As for cholesterol, one study found that supplements can contain as much as 600 milligrams of cholesterol per 100 grams.

- Do not supplement if you're pregnant or lactating, since high doses of fish oil may affect the health of your child.

- Do not take cod liver oil as a source of omega-3s; it is high in vitamin A and vitamin D, which in large amounts can be toxic.

- Fish oil supplements may increase your requirement for vitamin E, so ask your doctor if you would benefit from taking vitamin E supplements.

- Avoid fish oil supplements if you're taking blood-thinning medications (including aspirin), since the combination may increase your risk of bleeding.

- Avoid taking fish oil supplements two weeks before and one week after surgery, because these supplements can interfere with blood clotting.

- Some fish oil supplements may contain pesticides or other toxic contaminants.

- Discuss with your physician the advisability of taking fish oil supplements, especially if you are under medical care for a serious illness.

• F O U R •

DHA: The Brain-Building Fat

Though derogatory, the term fat head is a rather apt description of the brain. Fat makes up about 60 percent of your gray matter, and about a third of that fat is an omega-3 fatty acid called docosahexaenoic acid (DHA). There's even more DHA—50 to 60 percent—in the retina, a thin membrane attached to the back of your eye that senses and processes light images projected through your eyes.

Considered a building block of the brain, DHA is required for normal brain and eye development, as well as for mental well-being and visual functioning. In fact, your brain cells take up DHA in preference to other fatty acids.

DHA is also a constituent of cell membranes. One of its primary jobs is to protect the fluidity of brain cell membranes to ensure the normal transmission of nerve signals.

DHA is the most abundant omega-3 fatty acid in breast milk. Moreover, it is also found naturally in fish. No wonder, then, that fish is called a "brain food." Red meat, eggs, flaxseed, and certain vegetable oils contain appreciable amounts, too.

Deficiencies of DHA are linked to numerous health problems, including atherosclerosis, autoimmune diseases, arthritis, cancer, mental disorders, cancer, metabolic disorders, and nervous system problems.

DHA has been intensely researched in recent years and the findings are quite impressive. DHA has the following positive effects:

Banishes Bad Moods

Low levels of DHA are linked to depression, a mental illness that affects one of every four Americans. Case in point: A study published in the medical journal *Lancet* stated that in regions where people ate more fish, there were fewer cases of depression. What's more, researchers reported in the *American Journal of Clinical Nutrition* that the documented increase in depression in North America in the last century parallels the dwindling consumption of DHA over the same period.

In other research, scientists have examined the cell membranes of people suffering from depression to assess cellular DHA levels. One study of fifteen depressed patients and fifteen healthy volunteers found that the depressed patients had significant depletions of essential fatty acids, particularly DHA, in the cell membranes of red blood cells. Considering the evidence, many mental health experts are advocating DHA supplements as a natural treatment for mild to moderate depression.

Rejuvenates Your Brain

Some researchers speculate that DHA, because of its importance in human brain tissue, may help prevent degenerative brain diseases such as dementia, memory loss, and Alzheimer's disease. In fact, a study conducted at Tufts University discovered that a low level of DHA is a significant risk factor for these brain diseases. One reason for the shortfall may be due to the body's decreasing ability to synthesize DHA as we age. Another possible explanation, say scientists, is inadequate nutrition among the elderly. Research is now being conducted to learn whether supplementation with DHA can avert age-related declines in mental function.

Enhances Memory

DHA may bring about memory improvement, according to an animal study conducted at Shimane Medical University in Japan. Using rats bred on a fish oil–deficient diet, investigators fed DHA to one group of animals and a placebo to another for ten weeks. The goal of the study was to learn whether

DHA supplementation would have any effect on two types of memory: *reference memory*, which is information you learn and retain to perform a task the next time around; and *working memory*, another term for short-term memory.

DHA enhanced reference memory, but not short-term memory. The evidence: DHA-supplemented rats made fewer mistakes while negotiating a maze than the placebo-fed rats did.

Moreover, examination of the rats' brains revealed a possible biological clue for the improved memory: increased DHA in both the hippocampus, a sea horse–shaped structure deep within the brain that helps you learn and remember; and the cerebral cortex, an area of the brain associated with visual memory and recollection. Based on these observations, the investigators suggested that DHA levels in the brain may indeed influence learning ability.

Treats Attention-Deficit Hyperactivity Disorder (ADHD)

Some preliminary evidence indicates that less-than-adequate levels of DHA are correlated with behavioral problems in children. A Purdue University study found significantly lower levels of DHA in children with attention-deficit hyperactivity disorder (ADHD), compared to controls. Generally affecting children, ADHD is the habitual inability to pay attention for more than a few moments and is accompanied by erratic activity.

Boosts Kids' Brainpower

Children who were breastfed as infants generally have higher intelligence and greater academic achievement than infants who were formula-fed, according to studies spanning more than twenty years. Consider: One study found that breastfed babies have a 38 percent greater likelihood of completing their high school education than formula-fed babies.

Some researchers speculate that DHA, which is found naturally in breast milk, may be one of the reasons for the better-than-average mental functioning in children who were breastfed as newborns. Studies do show that breastfed infants have higher levels of DHA in their brains than do formula-fed babies, so there may be some connection.

Unlike in European and Asian countries, infant formula in the United

States is not enriched with DHA, so the only way babies can get this vital fatty acid is through breast milk. Unfortunately, though, breast milk levels of DHA in American moms are among the lowest in the world. For comparison, the DHA in the breast milk of Europeans is double that of American women, and the breast milk of Japanese women is three times as rich in DHA. Breast milk is considered the best source of DHA for infants.

Thus, in the United States, many health care practitioners recommend eating more fish if you're pregnant or lactating. Another natural food source of DHA is flaxseed oil. You can also take a daily DHA capsule, but consult your physician first.

Saves Your Infant's Sight—and Yours

Not only is DHA vital for an infant's brain development, it is also essential for normal visual development and healthy eyesight. As noted earlier, DHA accounts for more than a third of the fatty acids in the retina and strengthens brain cells associated with eyesight. In a mother's womb, DHA is taken up preferentially by the placenta and travels directly to the unborn baby's brain and retinal tissue. During the final months of pregnancy and within the first six months of infancy, the retina undergoes rapid development, drawing upon stored energy and large amounts of DHA for growth.

If an expectant mother's body is DHA-needy or her newborn does not get enough DHA during these critical stages of retinal development, the child's visual health may be compromised. Evidence for this comes from a couple of studies. In a study published in the *American Journal of Clinical Nutrition*, children who were breastfed for four months had better stereoscopic vision at age three and a half than children who were not breastfed. Stereoscopic vision is the ability to see objects in three dimensions. What's more, children whose mothers ate oily fish during their pregnancies had better stereoscopic vision than those whose mothers who avoided fish.

Similarly, a study from the Retina Foundation in Dallas, Texas, discovered that newborns fed formula without DHA for a year had poorer vision than babies fed breast milk.

Does DHA confer a vision-protecting effect in adults, too? Scientists think so. In a study reported in the *Archives of Ophthalmology*, investigators found that people who ate more fish, which is rich in DHA, had fewer incidences of age-related macular degeneration, the leading cause of legal blindness in the United States.

Treats Schizophrenia

Another potential use of DHA lies in treating schizophrenia, a destructive distortion of thinking in which a person's interpretation of reality is severely abnormal. Substantial evidence shows that chemical abnormalities in the brain cause schizophrenia.

A growing number of researchers are convinced that one of these abnormalities may be caused by low levels of DHA in brain cells and in red blood cells. British psychiatric researchers discovered that the depletion of essential fats, including DHA and arachidonic acid, is triggered largely by free-radical attacks on the membranes of red blood cells in patients with schizophrenia.

Boosting dietary levels of DHA may help, they found. After eating more essential fatty acids, including DHA, patients with schizophrenia had fewer symptoms after just six weeks.

Some startling research has revealed that newborns deprived of breast milk may be at risk of developing schizophrenia later in life. In a 1997 study published in the *British Journal of Psychiatry,* investigators suggest that a lack of brain-building DHA may contribute to this risk. They found that patients who were not breastfed had more schizophrenic traits and were more poorly adjusted than their breastfed siblings.

Protects the Heart

Enriching the diet with DHA lowers triglyceride levels by 26 percent, increases the good HDL cholesterol by 9 percent, and elevates apoprotein-E (a compound that ferries cholesterol from tissue back to the liver for breakdown and excretion) by 69 percent, according to a four-month study conducted by the U.S. Agricultural Research Service (ARS). That's encouraging news, particularly if you're trying to get your blood lipids under control.

Compared to fish oil supplements, isolated DHA seems to outperform on a number of fronts. For one thing, DHA does not result in such side effects as increased bleeding time or slower-than-normal blood clotting, both of which can occur with fish oil supplementation. Taking fish oil supplements also seems to elevate bad LDL cholesterol; taking DHA supplements does not.

Lowers Blood Pressure

Supplementing with DHA may help you defeat rising blood pressure. In a study conducted at the University of Western Australia, fifty-six men (ages twenty to sixty-five) took 4 grams of either DHA, EPA, or an olive oil placebo every day for six weeks. All the men were overweight, which is a risk factor for high blood pressure. Among the lipids, DHA was the only one that had any significant effect, lowering systolic pressure (upper number) by 3.5 points and diastolic pressure (bottom number) by 2 points during the day. The benefits were modest, but the researchers speculated that DHA may be the fatty acid in fish most responsible for regulating blood pressure.

Controls Blood Sugar

Research with lab animals hints that supplementing with DHA may put the brakes on a condition called *insulin resistance.* A hormone with multiple jobs in the body, insulin is required to move blood sugar (glucose) into cells for energy and nourishment. But in cases of faulty sugar metabolism, cells don't respond to insulin properly. Consequently, glucose is locked out of cells, and it clutters up the bloodstream.

The pancreas (where insulin is made) is forced to pump out more of the hormone to handle the excess glucose in the blood. But eventually, the pancreas can't keep pace with the demand, so there's a flood of both insulin and glucose in the bloodstream. This situation is a ticking time bomb that can explode into diabetes, and with diabetes, the potential and gradual deterioration of organs and blood vessels—unless blood sugar can be brought under control.

Researchers in Japan observed that DHA reduced glucose levels in DHA-supplemented mice—a finding they attribute to DHA's ability to make cells less resistant to insulin. In other words, DHA somehow altered cellular metabolism, allowing insulin to perform its job of pushing glucose into cells. Though far from conclusive, this study presents some intriguing evidence in favor of a diabetes-fighting benefit of DHA.

Supplementing with DHA

You can increase your supply of DHA by eating more cold-water fish (the current recommendation is two to three fish meals a week), incorporating a tablespoon a day of flaxseed oil into your diet, or taking DHA supplements. Also important: Limit or avoid excessive alcohol consumption, since chronic alcohol intoxication depletes DHA in brain cells.

Generally, DHA supplements are manufactured from fish oil. However, one product—Neuromins—is made from marine algae, the fish's original DHA source, through a special extraction process. This product is free of toxins that may be present in some fish oils.

The recommended dosage of DHA is 100 milligrams a day for healthy adults who obtain some DHA from fish and other sources. If you eat little or no fish, 200 milligrams a day is recommended.

DHA is a fatty acid that is naturally present in various foods, and so it is digested just like any other fat. It appears to be very safe, with virtually no known side effects. As with any dietary supplement, you should consult a health care professional before taking supplemental DHA.

• F I V E •

Nature's Disease Fighters

Two good fats that could upstage a headliner like fish oil are now—bit players called flaxseed oil and perilla oil. Both are highly concentrated sources of the omega-3 fat alpha-linolenic acid, which has a windfall of health benefits. If you started eating both of these oils on a regular basis, you'd probably escape such feared diseases as heart disease, stroke, cancer, and arthritis, to name just a few.

Just the Flax, Ma'am!

For background, flaxseed oil comes from the crushed seeds of the flax plant, a blue flowering plant cultivated in Europe, South America, Asia, Canada, and parts of the United States. Historically, the flax plant has been used as medicine since at least 3000 B.C. Later, in 650 B.C., Hippocrates, "the Father of Medicine," prescribed flaxseed as a curative for intestinal problems. The eighth-century king Charlemagne issued laws governing and protecting the plant, so important was flaxseed for the health of his subjects.

Flaxseed is a widely recognized folk remedy throughout the world, and today many health-conscious consumers are eating it for its many healthful benefits. Flaxseed is officially approved by the German Commission E—Germany's version of our FDA—for treating constipation, diverticulitis, irritable bowel syndrome, and colons damaged by laxative abuse.

Worth mentioning too: On many farms, chickens are fed rations containing flaxseed so that they will lay omega-3-enriched eggs. In fact, these eggs contain eight to ten times more omega-3 fatty acids than regular eggs.

Though flaxseed is not a universal cure-all, this much is certain: Flaxseed and flaxseed oil display astonishing powers against many life-shortening diseases. This power emanates from the various constituents of flaxseed. Flaxseed oil is simply the world's richest treasure trove of omega-3 fatty acids. In fact, it has more than double the amount of omega-3s found in fish. So if you're not a fish eater, you can still get plenty of omega-3s by eating flaxseed and flaxseed oil.

Flaxseed is also a super source of alpha-linolenic acid (ALA), yielding 57 percent ALA—compared to 10 percent ALA found in the next-highest sources, canola and walnut oils. Flaxseed oil also contains 16 percent linoleic acid (an omega-6 fatty acid), 18 percent monounsaturated fatty acids, and only 9 percent saturated fatty acids.

Another ingredient in flaxseed worth knowing about is a group of beneficial phytochemicals (plant chemicals) called *lignans,* which work in monumentally important ways. First, they are thought to act as antioxidants, saving cells from destructive free radicals. Second, they also function as phytoestrogens, weak versions of estrogen found in fruits, vegetables, and whole grains.

Phytoestrogens are therapeutic in that they have a split personality. In premenopausal women, who have a lot of circulating estrogen, phytoestrogens cause the body to produce less of the hormone. Yet they work just the opposite way in postmenopausal women, who have low levels of the hormone, by increasing levels of estrogen. Many health practitioners believe that the estrogen-regulating effects of phytoestrogens may be helpful in fighting hormone-dependent cancers, such as breast or uterine cancers. Indeed, women who eat a lot of lignans have lower rates of breast cancer, according to research.

Lignans are found in numerous plant foods—namely barley, buckwheat, millet, oats, legumes, vegetables, and fruits—but nowhere are they as abundant as in flaxseed. Flaxseed boasts 75 to 800 times more lignans than any other food in the plant kingdom. Flaxseed oil, however, contains negligible amounts of lignans, which are nearly all removed during the oil-extraction process.

Flaxseed is loaded with fiber too, with about 6 grams in a fourth of a cup—more than twice as much as you'd get in a comparable serving of pure wheat bran. Research shows that adding flaxseed to your diet promotes regularity.

If you're really serious about enhancing your health, it's time to get serious about flaxseed. Here is a more detailed look at its health-giving properties.

Unclogs Arteries

Eating about 3½ tablespoons of milled flaxseed daily (in muffins) reduces artery-clogging LDL cholesterol by 18 percent and total cholesterol by 9 percent, a Canadian study found. Animal studies indicate that the lignans in flaxseed are responsible for the cholesterol-lowering effect.

Scientists believe that lignans, which are antioxidants, suppress the activity of free radicals before they have a chance to oxidize cholesterol and damage the interior linings of the arteries. Investigators at the University of Saskatchewan, who have conducted the bulk of this research, note that supplementing with flaxseed could "prevent hypercholesterolemia-related heart attack and strokes." Hypercholesterolemia is the medical term for high cholesterol.

Further, postmenopausal women who ate bread and muffins containing 38 grams of flaxseed flour lowered their total cholesterol from 229 to 213 and their LDL from 158 to 133, on average. Significantly, flaxseed reduced levels of a protein called *lipoprotein a,* which increases after menopause and contributes to the development of atherosclerosis, the narrowing and thickening of arteries that is responsible for heart attacks.

Prevents Second Heart Attacks

As stated earlier, flaxseed is loaded with alpha-linolenic acid (ALA). In a five-year study of survivors of a first heart attack, researchers found that those who followed an ALA-enriched diet, compared to those who followed a standard low-fat diet, had fewer second heart attacks over the experimental period. The researchers noted that the ALA-enriched diet appeared to be effective in preventing second heart attacks, but added that more research is needed to confirm a benefit.

Normalizes Clotting Function

In a twenty-three-day study involving college-age men, flaxseed oil (40 grams daily) exhibited "antiplatelet" activity. This means it prevents platelets from clumping together when they're not supposed to in an abnormal process called platelet aggregation. Platelet aggregation triggers the formation of dangerous blood clots called *thrombi* (singular, *thrombus*). A heart attack occurs when a thrombus obstructs a blood vessel supplying oxygen and nutrients to the heart muscle. The researchers concluded that alpha-linolenic-rich oils such as flaxseed may offer protection against cardiovascular disease complicated by the potential for blood clots.

Thwarts Cancer

Flaxseed displays cancer-fighting power, possibly due to its abundance of lignans. As noted, lignans are phytoestrogens that may protect against certain kinds of cancer, particularly hormone-dependent cancers such as those of the breast and prostate.

Studies at the University of Toronto showed that daily doses of flaxseed shrank mammary tumors in rats by more than 50 percent after seven weeks of treatment. Again, scientists believe that the tumor-reducing effect is due to lignans. So promising is the animal research that flaxseed is being tested to shrink breast tumors prior to surgery in women with breast cancer. Lignans have also been shown in animal studies to keep new tumors from sprouting.

Other animal studies show that flaxseed protects against colon cancer—a benefit that is also linked to lignans—and prevents the spread (metastasis) of this deadly cancer.

Boosts Immunity

The omega-3 fatty acids in flaxseed favorably affect your immune system by beefing up the content of alpha-linolenic acid, docosahexaenoic acid (DHA), and eicosapentaenoic acid (EPA) in the membranes of immune cells. Increasing the omega-3 content of cell membranes does two important duties. First, it helps encourage the formation of beneficial prostaglandins and leukotrienes. Second, it inhibits the formation of proteins that promote inflammation, encourage cancer, and interfere with normal cell function. One study showed that eating a flaxseed oil–enriched diet for eight weeks suppressed the production of these harmful proteins by 77 to 81 percent in a group of men. Bottom line: Flaxseed appears to inhibit harmful reactions leading to cell damage.

Prevents Diabetes

Diabetes is a sugar metabolism disorder in which the body does not produce any insulin (type I diabetes), or can't use it properly (type II diabetes). Some diabetes investigators believe that free-radical damage may contribute to diabetes by damaging insulin-producing cells in the pancreas. In type I diabetes, especially, cells in the pancreas are naturally low in levels of protective antioxidants that neutralize free radicals.

Now it appears that adding flaxseed to your diet may be a good defense against free-radical damage to the pancreas, according to a Canadian study. The study found that rats treated with a lignan isolated from flaxseed were protected from developing type I diabetes. According to investigators, the lignan shored up antioxidant defenses and prevented *oxidative stress*—a situation that occurs when protective antioxidants are swamped by destructive free radicals. Flaxseed oil may also boost blood circulation, which is often sluggish in people with diabetes.

Flaxseed Oil vs. Fish Oil

There's no question that supplementation with fish oil confers some demonstrable health benefits, particularly in normalizing cholesterol and blood pressure. But so does flaxseed oil. The reason for their therapeutic power is that both oils are rich in omega-3 fatty acids. So the question is: Which oil is better if you want to acquire the healing power of omega-3 fatty acids?

Here are a few things to consider. Some fish oil supplements have been found to be adulterated with high levels of *lipid peroxides*. These harmful byproducts form when free radicals hook up with fatty acids. Lipid peroxides attack cell membranes, setting off a chain reaction that creates many more free radicals. Pits form in cell membranes, allowing harmful bacteria, viruses, and other disease-causing agents to gain entry into cells. Thus, a better strategy may be to eat more fish and supplement your diet with flaxseed oil, rather than rely on fish oil supplements for your omega-3 fatty acids.

Further, flaxseed oil:

- Has less tendency to turn rancid

- Is not contaminated with toxic substances (fish oil supplements are rarely pure and often contain concentrations of highly toxic chemicals)

- Is more economical (at recommended dosages, fish oil can cost an average of $40 to $100 a month, while flaxseed oil averages between $6 and $18 a month)

Purchasing Quality Flaxseed Oil

There are many manufacturing variations in the way flaxseed oil is made from seeds, but generally it is produced by mechanically extracting and pressing the oil through an expeller press. The process generates a tremendous amount of heat, which can easily damage the oil's essential fatty acids. The higher the temperature, the greater the yield. Unfortunately, many manufacturers sacrifice quality for quantity.

Because of the variations in processing, not all flaxseed oil has the same quality. Nor does it possess the same therapeutic potential. Thus, when shopping for flaxseed, purchase the best-quality oil you can find. To help you, here are some guidelines recommended by herbalists and alternative health practitioners:

- Purchase flaxseed oil from a health food store. Generally, health food stores carry brands from quality manufacturers. In addition, make sure the oil is found in the refrigerated case of the store. Storage at room temperature for prolonged periods can degrade the potency of the oil.

- Check to see whether the oil is produced by *modified expeller presses* at temperatures that do not exceed 98 degrees Fahrenheit. Some of the trade names to look for are Bio-Electron Process, Spectra-Vac, and Omegaflo. Do not purchase oil that has been *cold pressed*. This is a very deceiving term because processing temperatures can reach 200 degrees Fahrenheit, even though no external source of heat is used.

- Make sure the product is packaged in an opaque (light-resistant) container. Contact with light can degrade the oil and destroy its healthful properties.

- Purchase an oil that is certified as organic by a third-party source. Stated on the label, certification ensures that the oil is free from pesticides and herbicides.

- Check the expiration date of the oil and adhere to these dates. You can freeze flaxseed oil, however, and thus extend its expiration date indefinitely.

Using Flaxseeds and Flaxseed Oil

The two best ways to get the benefits of flaxseed are to add whole flaxseeds or flaxseed oil to your diet. For good health, the recommended dosage is 1 tablespoon of whole seeds for every 100 pounds of body weight. You can sprinkle flaxseeds on your cereal in the morning—they have a delicious nutty flavor—put them in salads, mix them into yogurt or applesauce, or bake them into muffins and breads in place of nuts. The alpha-linolenic acid and lignans in flaxseeds remain stable when batters and doughs containing them are baked at customary baking temperatures.

You can also mill the seeds in a coffee grinder to yield a finer consistency. Ground seeds can be added to various recipes to enhance the flavor and nutrition of foods. Because of their rich oil content, ground seeds can replace all of the oil or shortening in recipes. If a recipe calls for one-third cup of oil, for example, use 1 cup of ground seeds instead (a 3-to-1 substitution). Also, one ½ cup of ground flaxseed can replace ½ cup of butter, margarine, or shortening. Baked goods using seeds in this manner will brown more rapidly.

You can also substitute ground flaxseed for eggs in recipes if you are a vegetarian or are cutting back on eggs. The substitution is as follows: 1 tablespoon ground flaxseed plus 3 tablespoons of water=1 egg. Let the flaxseed/water mixture sit for 1 to 2 minutes before using in the recipe.

For treating gastrointestinal problems, *The PDR for Herbal Medicines* suggests taking 1 tablespoon of whole seeds with a half cup of liquid two to three times a day. Whole seeds should be stored in a cool, dry place and can be kept for one year. In addition, shop for flax-enriched foods, which are turning up in health food stores and some grocery stores.

Another option is to consume 1 tablespoon of flaxseed oil daily, in the form of a flaxseed salad dressing or mixed into food. Flaxseed oil is digested and absorbed better when eaten with food. Recipes for using flaxseeds and flaxseed oil are found in chapter 12.

Flaxseed oil tends to turn rancid very quickly, so it should be stored in your refrigerator. You can add the oil to cooked food, but do not use it during cooking because heat destroys its beneficial properties. Flaxseeds can be stored at room temperature for up to one year; ground seeds should be stored in your refrigerator in an airtight, opaque container.

Flaxseed oil also comes in capsules, but they are expensive compared to buying the bulk oil. Some health care practitioners and herbalists believe that flaxseed oil capsules, though convenient, are not as high quality as the pure oil because of the manufacturing process used to encapsulate them.

Side Effects

No significant side effects or health hazards have been associated with consuming flaxseeds or flaxseed oil at the recommended therapeutic dosages. The absorption of drugs taken simultaneously with flaxseed may be delayed, so you should consult your health care practitioner or pharmacist about possible drug interactions if you are taking prescription medicine.

Perilla Oil—An Up-and-Coming Good Fat

This rather obscure oil is derived from the seeds of the Asian beefsteak plant (*Perilla frutescens*), cultivated in Asian countries. Its leaves are used as a spice and in pickled food in Japan. No wonder, then, that most of the scientific research on perilla oil has been conducted in Japan. You can purchase perilla food products in the United States in many Korean grocery stores. Supplements made from the oil are available commercially in health food stores.

Perilla oil is packed with good fats and may even rival flaxseed oil in terms of its alpha-linolenic acid content, according to some analyses, making perilla oil one of nature's richest sources of omega-3s. The oil is also high in several *phenolic compounds*, natural chemicals that are therapeutically beneficial to health.

In Asia, herbalists prescribe perilla for relieving coughs, enhancing lung health, and fighting colds and flu. Various health claims have been attached to perilla oil, and these include its potential ability to fight allergies and asthma, ease pain and inflammation, stimulate immunity, and guard heart health. Some of these have been studied in research; others have not.

Whether perilla oil will be as beneficial as other healthful oils is not yet known, since very few clinical trials have been conducted with humans. Most of the research has been done in animals. Nonetheless, here's a look at what is currently known about this oil.

Treats Asthma

Affecting nearly 10 million Americans—most of them children—asthma is a serious, sometimes fatal disease. Its primary symptom is difficulty in breathing. During an asthma attack, antibodies react with allergens, producing histamine and other chemicals, which are responsible for most of the symptoms of asthma. Among the chemicals generated are pro-inflammatory leukotrienes.

In most cases, asthma is treated with inhaled medicine, oral doses of steroids and other drugs, and avoidance of allergens and pollutants. Recently though, perilla oil has been investigated as a potential treatment for asthma.

In one study, fourteen patients with asthma were randomly divided into two groups. For four weeks, one group supplemented with perilla oil; the other, with corn oil supplements. The findings: In the perilla oil–supplemented group, the generation of leukotrienes and other asthma-provoking substances was suppressed, whereas in the corn oil group, these chemicals increased. Lung function also improved in patients taking perilla oil. The researchers noted that perilla oil is useful for the treatment of asthma by suppressing substances that aggravate the disease and by strengthening breathing capacity.

Saves Hearts

Evidence is popping up that perilla oil may be good for your heart, according to researchers in Japan. They tested the effects of replacing soybean cooking oil with perilla oil (providing 3 grams a day of alpha-linolenic acid) in the diets of twenty elderly patients. The experiment worked like this: For six months, the subjects ate a diet enriched with soybean oil. After that, they consumed a diet enriched with perilla oil. Finally, they were switched back to the soybean oil–enriched diet so that researchers could compare the effects of the diets. The major finding of this study was that the 3-gram-a-day increase in alpha-linolenic acid caused the patients' bodies to churn out more EPA and DHA, both of which are heart-protective.

Perilla oil has also been shown to decrease abnormal blood clotting by 50 percent in studies of rodents. Lodged in an artery, a blood clot can narrow the passageway and choke off blood flow, leading to heart attack or stroke. Perilla oil appears to work in two ways: by reducing the production of platelet activating factor, a chemical released by cells that causes platelets (clotting elements in blood) to stick together; and by suppressing the production of thromboxane, another substance that makes platelets clump together and more apt to form clots.

Wards Off Cancer

Another potential plus for perilla oil: It may reduce the risk of colon cancer. In a Japanese experiment involving rats, researchers discovered that a small amount of perilla oil added to the diet (25 percent of total dietary fat) reduced the overall incidence of colon cancer in the animals. This led the researchers to note that perilla oil may help lower the risk of colon cancer.

In other animal studies, perilla oil has demonstrated an antitumor effect against tumors of the mammary gland and kidney.

Controls Weight

It may be just a blip on the research radar, but investigators in Japan discovered that perilla oil can shrink fat cells, prevent fat cells from enlarging (which results in obesity), and reduce the weight of fat pads in rats. The oil can also significantly reduce concentrations of triglycerides, or blood fats, in the animals. In their report, published in the *Journal of Nutrition*, the researchers wrote: "Therefore, daily consumption of perilla oil by humans seems advantageous."

It's certainly too early to tell whether perilla oil is a bona fide fat burner. However, these researchers are planning to study the antifat effect of perilla oil in humans—so stay tuned.

Supplementing with Perilla Oil

Perilla oil is available in supplements, and the typical daily dosage is around 6 grams, which provides 3 grams of alpha-linolenic acid. Dosage recommendations may vary from manufacturer to manufacturer, so read product labels for instructions. Check with your health care professional prior to supplementing with perilla oil.

Side Effects

Most studies with perilla oil have been conducted with animals, so there is little information on its long-term safety in humans. Even so, perilla oil is widely consumed in a variety of Asian food products and has not been linked to any harmful side effects. Some health care practitioners recommend perilla oil over fish or flaxseed oil supplements because it is reportedly gentler on the stomach.

• SIX •

Omega-6 Healers

Although you should strive to pump up your omega-3 intake, you still need omega-6 fatty acids in your diet to help regulate and balance body processes. The key is to select better-quality omega-6 fats over inferior versions (generally found in margarine and processed foods). Higher-quality omega-6 oils include borage oil, black currant seed oil, and evening primrose oil, all of which can be used therapeutically to treat disease and strengthen health.

From this trio of oils, your body can directly obtain gamma-linolenic acid (GLA). In the body, GLA is also synthesized from the essential omega-6 fatty acid linoleic acid. GLA is ultimately metabolized into the prostaglandin E1 series, a group of beneficial chemicals that helps reduce inflammation, regulates blood clotting, decreases cholesterol levels, and lowers high blood pressure, among other functions.

Other omega-6 oils—namely sesame oil, wheat germ oil, and hempseed oil—are marketed as nutritional supplements. Some of these oils contain antioxidants and other substances beneficial to health. Here's a closer look at all of these omega-6 healers.

Borage: Nature's All-Star Oil

Some history: Borage is a plant that was originally cultivated in Syria, and its name means "father of sweat" in Arabic, a reference to its use as a diaphoretic (an agent that increases perspiration). The ancient Romans were fond of borage, and Pliny the Elder recommended steeping it in wine to make a person more merry.

In folk medicine, borage has been used as a treatment for coughs and throat illnesses; as an anti-inflammatory agent for kidney, bladder, and joint problems; as a pain reliever; and as treatment for menopausal complaints.

Borage oil comes from the seeds of the borage plant and is a highly concentrated source of GLA, boasting 22 percent. A number of health conditions can benefit from treatment with GLA-rich borage oil.

Reduces Blood Pressure

A contributing factor to high blood pressure (hypertension) is stress in our lives. Happily, supplementing with borage oil may help you reduce stress-related high blood pressure, says a Canadian study. Researchers randomly assigned thirty men to one of three groups—borage oil supplements, fish oil supplements, and olive oil supplements—to see which supplements could reduce high blood pressure, provoked in response to a special type of test designed to induce stress. Borage oil alone kept blood pressure down, encouraged normal heart rate, and improved the subjects' performance on the test. Animal research also indicates that borage oil has a blood pressure–lowering effect.

Treats Respiratory Distress

Borage oil may save the lives of patients at risk for acute respiratory distress syndrome (ARDS). A common problem among patients in hospital intensive care units, ARDS kills approximately 50 percent of sufferers. The condition is associated with extensive lung inflammation and blood vessel injury in affected organs. ARDS comes on suddenly, and the lungs fill up with liquid. Patients drown in their own secretions.

Those at risk include trauma victims and patients with sepsis (internal infection), pneumonia, or shock, according to the American Lung Association. ARDS is also linked to extensive surgery, drowning, and inhalation of toxic gases. Conventional treatment consists of mechanical ventilation along with fluid removal and a special type of breathing technique.

A study of 150 at-risk patients at the Mayo Clinic found that borage oil reduced the mortality rate by 36 percent. Fluid samples taken from patients' lungs before and during treatment showed significantly lower white blood cell counts in those who were treated with borage oil, indicating reduced inflammation. The reason for the oil's healing effect: GLA, which reduces inflammation and improves oxygen flow in the body.

More studies are being conducted to confirm what doctors are beginning to suspect: that borage oil may improve the chances of recovery from ARDS.

Eases Rheumatoid Arthritis

More than 2 million people suffer from rheumatoid arthritis, a crippling autoimmune disease that strikes people of all ages, mostly women. Its symptoms include fever, joint stiffness and swelling, fatigue, muscle weakness, loss of appetite, and depression. Rheumatoid arthritis is "symmetrical," too, affecting the same joints on both sides of the body.

In rheumatoid arthritis, the body overproduces inflammation-producing chemicals, leading to the painful symptoms of the disease. Of all the plant oils, borage oil seems to work the best at soothing inflammation in this debilitating disease.

In a twenty-four-week clinical trial at the University of Pennsylvania, researchers treated thirty-seven rheumatoid arthritis patients with 1.4 grams a day of borage oil or cottonseed oil (the placebo). Treatment with borage oil produced some near-miraculous results: It reduced the number of tender joints by 36 percent and swollen joints by 28 percent. No such improvements were reported in the placebo takers. These findings led the researchers to state that borage oil was "a well-tolerated and effective treatment for active rheumatoid arthritis."

Borage oil may also benefit children with mild juvenile rheumatoid arthritis. In one study, children who took approximately 1 to 2 teaspoons of borage oil daily for a year experienced improved joint mobility, less morning stiffness, and fewer tender or swollen joints. They supplemented with borage oil in addition to their regular medicine, leading doctors to conclude that the oil may serve as an effective complementary treatment for juvenile rheumatoid arthritis.

Using Borage Oil

Herbalists suggest taking two to eight supplements of borage oil daily for its therapeutic benefits, but no more than 3 grams a day. You can also make a tea from the dried leaves of the borage plant. Add a cup of boiling water to about 2 teaspoons of dried leaves.

Borage oil should not be taken if you are taking anticonvulsants because it may interfere with the action of these drugs.

Borage oil is not appropriate for cooking because heat breaks down its chemical structure, and its fatty acids are converted to toxic by-products called lipid peroxides. Borage oil is best used as a medicinal oil.

It is important to add that borage is not approved by the German Commission E. One reason is that borage contains small amounts of plant chemicals called pyrrolizidine alkaloids (PA), which can contribute to liver damage and have been shown to be cancer-promoting in animal tests. That being the case, discuss the advisability of supplementation with your health care practitioner.

Black Currant Seed Oil: Prescription for Natural Healing

Black currant seed oil, derived from the seeds of the Ribes nigrum plant grown in swamps and rain forests, contains roughly 17 percent GLA and is one of nature's richest sources of this fatty acid. It is also a good source of the omega-3 fat alpha-linolenic acid, and thus provides a good balance of omega-6 and omega-3 fats.

What sets this oil apart is that it is the only commercially available oil endowed with a fatty acid called *stearidonic acid,* a by-product of alpha-linolenic acid. Investigators speculate that stearidonic acid is behind many of the health benefits associated with black currant seed oil.

Black currant seed oil has not been extensively researched; nonetheless, there is some evidence that it confers some important disease-fighting benefits.

Boosts Immunity

Supplementing with black currant seed oil may stimulate your immune system, which tends to weaken with age. In a study conducted at Tufts University, twenty-nine people older than age sixty-five took 4.5 grams of black currant seed oil or soybean oil (the placebo) every day for two months. Researchers checked their immune function by measuring their skin response to a toxin. Response to this particular toxin generally declines with age.

By the end of the experimental period, those who supplemented with black currant seed oil showed a 28 percent greater skin response than those on the placebo—which meant that their immune systems had rallied to fight off the toxin.

The researchers attributed the oil's immune-bolstering effect to the ability of GLA to suppress the production of prostaglandin E2, known to have an adverse effect on immune factors. Also, because older adults produce less GLA, they may have benefited from supplementation with a natural source of GLA.

Treats Rheumatoid Arthritis

Black currant seed oil is another fat that may be helpful in easing symptoms of rheumatoid arthritis. In one study, patients taking 525 milligrams of the oil every day experienced less morning stiffness.

Relieves Vaginal Dryness

Vaginal dryness occurs with age as vaginal tissues thin out, become less elastic, and stop secreting as much moisture. Herbalists suggest supplementing with black currant seed oil or other GLA-rich oils. The oil is believed to be important for the production of estrogen. Declining estrogen levels contribute to vaginal dryness.

Other Benefits

Research into the health benefits of black currant seed oil is under way, and preliminary findings, mostly with animals, show that it may also:

- Relieve diabetic nerve damage, a serious complication of diabetes
- Reduce high blood pressure
- Inhibit unhealthy blood clotting
- Suppress tumor growth

Using Black Currant Seed Oil

The recommended dosage of black currant seed oil ranges from two to eight capsules a day. Many health practitioners recommend taking no more than 4 grams a day.

Although there are no known side effects associated with taking black currant seed oil, check with your health care practitioner before deciding to supplement.

Evening Primrose Oil: Nature's Answer for What Ails You

If your joints are creaky, your skin is dry and lifeless, or your menstrual symptoms are harder than ever to endure, don't despair. Reach for a gentle-on-your-body natural remedy called evening primrose oil, used down through the ages to treat a wide range of ailments.

Evening primrose oil comes from a plant that grows wild along roadsides. It is so named because its yellow flowers resemble real primroses, and these flowers open only in the evening.

The oil extracted from the seeds of this plant yields roughly 9 percent GLA, along with appreciable amounts of alpha-linolenic acid. Evening primrose oil is indicated for any diseases or conditions with which prostaglandins are associated, and these include premenstrual syndrome (PMS); heart disease; diabetic neuropathy, a type of nerve damage that is a complication of diabetes; and arthritis.

Evening primrose oil has long been used in Britain for treating women's health problems, particularly PMS. Here is a rundown of specific conditions for which evening primrose oil may be beneficial.

Eases Menstrual Symptoms

Among American Indian women, chewing the oil-rich seeds of the evening primrose plant is a centuries-old treatment for premenstrual and menstrual problems. And when 300 Australian women were asked to name their favorite treatments for PMS, the top three were dietary changes, vitamins, and evening primrose oil. What's more, evening primrose oil is an approved treatment for PMS in Great Britain.

And no wonder. This rather remarkable oil is an excellent alternative treatment for alleviating such symptoms as fluid retention, headaches and backaches, skin problems, food cravings, depression, tension, irritability, fatigue, weeping, tantrums, and lack of concentration.

Banishes Breast Pain

Supplementing with evening primrose oil may be the best way to ease breast pain (mastalgia) during menstruation. Painful breasts are a common menstrual and premenstrual problem that can range in severity from mild to severe pain. Mild breast pain, in fact, is considered normal and subsides when the woman's period is over.

Rest assured that breast pain is rarely a sign of cancer. Breast cancer–related pain tends to occur only in very-late-stage breast cancer, many years after treatment has started.

When breast pain is so severe that it interferes with daily living, many women often resort to prescription medications, either danazol (Danocrine) or bromocriptine. However, the side effects of these drugs can be worse than the breast pain itself. If you need relief, try evening primrose oil first. In studies, it has proven as effective as these drugs, but with no side effects. One study found that evening primrose oil was 97 percent effective against breast pain.

The reason it works so well is that many women with breast pain may be deficient in PGE1. A short supply of this prostaglandin intensifies the pain-inducing effect of another hormone, prolactin, on breast tissue. Evening primrose oil supplies extra GLA, which is ultimately converted to pain-relieving PGE1.

Acts as an Antioxidant

Among oilseeds, evening primrose has demonstrated some of the most powerful antioxidant activity. In one study, evening primrose oil snuffed out the hydroxyl radical, a form of hydrogen peroxide. Once generated, this devilish free radical attacks whatever is next to it, setting off a dangerous chain reaction that creates many more free radicals. In the same study, evening primrose reduced the formation of "superoxide" free radicals—a type of free radical that is particularly harmful to heart cells.

Cleans Out Arteries

Taking evening primrose oil may be one more weapon in your natural arsenal against high cholesterol. In a study of ten patients with high cholesterol, supplementation with 3.6 grams a day of evening primrose oil significantly reduced artery-clogging LDL cholesterol by 9 percent. Further, animal experiments show that evening primrose oil reduces plaque, which consists of fatty deposits that accumulate in arteries, narrow their passageways, and choke off blood flow.

Treats Diabetic Nerve Disease

Evening primrose oil has long been used as an alternative treatment for diabetes. One reason is that people with diabetes are often deficient in a special enzyme called delta-6-saturase, which is responsible for building many important components in the body. Without this enzyme, the body cannot convert the essential fatty acid linoleic acid into GLA. The impairment of such a critical enzyme creates a deficiency in essential fatty acid metabolism, which in turn leads to diabetic complications, primarily poor nerve conduction and sluggish blood flow in the extremities.

As noted earlier, evening primrose oil contains GLA. That's a plus because GLA does not require delta-6-saturase for breakdown by the body. Thus, some alternative health care practitioners believe that supplementation with evening primrose oil can supply needed amounts of GLA, circumvent an essential fatty acid deficiency, and thus help prevent diabetic complications.

Most of the research supporting this view has been conducted with animals, but the results are promising nonetheless. Nearly all of the studies have found that evening primrose oil corrects impaired nerve conduction and poor blood flow. In Germany, evening primrose oil is among the most widely recommended treatments for diabetic neuropathy.

Soothes Eczema

Because it lubricates your skin, evening primrose oil is an effective natural treatment for eczema, an inflammation of the skin that produces dryness, scaling, flaking, and itching. In clinical trials, evening primrose oil has been shown to reduce the severity of eczema and keep the skin smooth. As a result, many alternative health care practitioners recommend taking 500 milligrams of evening primrose oil each day to treat eczema. You should notice an improvement in your skin in about six to eight weeks.

The skin-smoothing benefits of evening primrose oil can be credited to GLA, which also has a hand in promoting healthy skin, hair, and nails. Scientists believe that the reason some people are prone to eczema is a defect in the enzyme delta-6-saturase, required for the conversion of linoleic acid to GLA. The body doesn't convert linoleic acid properly, resulting in a shortfall of GLA.

As explained before, supplementing with GLA-rich evening primrose oil circumvents this deficiency because GLA does not require the delta-6-saturase enzyme for breakdown by the body.

Combats Joint Problems

A growing number of medical experts and scientists now believe that taking GLA-rich oils such as evening primrose oil, borage oil, or black currant oil can effectively fight the inflammation—the major cause of swollen, painful joints—that is so characteristic of arthritis. As explained before GLA is a building block of the beneficial prostaglandin PGE1, which has an anti-inflammatory effect on the body. Thus, supplementing with GLA increases production of these prostaglandins and may help control the pain and inflammation associated with arthritis.

Fights Obesity

Evening primrose oil may be an antifat agent, too. GLA, one of its beneficial constituents, has been researched for its involvement in weight loss in both animals and humans. In studies with rats, GLA reduced body fat content.

As for humans, some scientists believe that people with GLA deficiencies tend to produce more fat in their bodies. Supplementing with evening primrose oil has helped them lose weight.

Research has found that the supplement works best if you are more than 10 percent above your ideal weight. And, in some people, evening primrose oil promotes weight loss without reducing caloric intake. It is also believed to help rev up the metabolism so that you burn more calories.

Evening primrose oil helps reduce fluid retention, medically known as *edema*. When you retain water, you look and feel "fat," even though you aren't. Encouraging the body to get rid of extra water can make you look more trim. One way to do this is by supplementing with evening primrose oil, which can help your body regulate water more normally and prevent fluid retention.

Because of its potential antifat and anti-edema properties, evening primrose oil, along with grape-seed oil (another oil rich in omega-6 fats), is an ingredient in supplements designed to minimize cellulite. Cellulite is a cosmetic condition related to the deteriorating underlying structure of the skin. It is characterized by a lax, dimpled skin surface covering the thighs, buttocks, and hips. Grape-seed oil in particular may help improve circulation, strengthen connective tissue, and reduce fluid accumulation—all factors that improve the appearance of cellulite-ridden skin.

Kills Tumor Cells

In 1992, a group of researchers in India observed something quite exciting: that evening primrose oil killed tumor cells isolated in lab dishes. Why did the oil act like such a powerful terminator? The scientists chalked the effect up to the presence of linoleic acid and GLA in the oil, which may be able to slow down cancer.

In an animal study conducted by Welsh researchers, rats given evening primrose oil, fish oil, or a placebo oil were studied to determine the effect of these various oils on breast tumor growth. Breast tumors were significantly smaller in those animals fed evening primrose oil and fish oil.

Keep in mind that these experiments are not proof of an antitumor effect of evening primrose oil, only preliminary evidence. Without question, more studies are needed in this promising area of research.

Manages Migraines

More than 28 million Americans suffer from migraines—painful, often debilitating headaches that can last from hours to days or even weeks at a time. Migraine pain throbs, usually on one side of the head, or feels like a hammer pounding your head. Headaches are often accompanied by nausea and vomiting, as well as sensitivity to light and sound. About 20 percent of sufferers experience visual disturbances or auras prior to the actual migraine attack.

Headache specialists agree that migraine is a disorder of the central nervous system, but one that is not well understood. One theory holds that when something triggers a migraine, blood vessels in the head abnormally constrict and dilate, causing pain and inflammation. For this reason, herbalists recommend evening primrose oil to treat migraines because it is an anti-inflammatory agent that keeps blood vessels from constricting.

Some researchers believe that migraines can be alleviated by tinkering with essential fatty acid content in the diet. In a study conducted at the University of Berlin, 168 migraine sufferers supplemented with essential fatty acids for six months. The supplements contained 1,800 milligrams of GLA and alpha-linolenic acid. The patients were told to avoid high doses of arachidonic acid, found mostly in omega-6 fats. By the end of the study, 86 percent of the patients experienced a reduction in the severity, frequency, and duration of their migraine attacks, 90 percent had fewer incidents of nausea and vomiting, and 22 percent became migraine-free.

The researchers felt that the improvements came about in the following way: From both GLA and alpha-linolenic acid come inflammation-reducing prostaglandins and leukotrienes, and from arachidonic acid come prostaglandins and leukotrienes that step up inflammation. So by jacking up dietary levels of GLA and alpha-linolenic acid and cutting back on sources of arachidonic acid, the good prostaglandins and leukotrienes dominated in cells in order to produce an anti-inflammatory effect.

Using Evening Primrose Oil

No one knows yet what dosage of evening primrose oil is most beneficial. However, most alternative health care practitioners recommend a dosage of 1 to 4 grams a day, in divided doses, for general well-being. If you're treating breast pain, you may want to increase the dosage to 6 grams daily.

Evening primrose oil comes in softgel capsules. Typically, each capsule contains 500 milligrams of the oil and supplies 5 calories per capsule. Make sure the product supplies at least 45 milligrams (9 percent) of GLA per capsule.

Evening primrose oil appears to be safe, with very few side effects. Some potential side effects include stomach upset and headaches. As with any supplement, do not take evening primrose oil if you're pregnant or lactating.

Sesame Oil

The seeds of the sesame plant yield an abundant supply of omega-6 fats, disease-fighting antioxidants, and other therapeutic chemicals. A major component of the oil, for example, is an antioxidant called *sesaminol*. In experiments conducted in Japan, investigators observed that sesaminol worked better than vitamin E at guarding LDL cholesterol against free-radical–generated oxidation, which leads to artery damage. Based on this finding, they noted that "sesaminol is a potentially effective antioxidant that can protect LDL against oxidation." Sesame seeds also are rich in cholesterol-lowering compounds called *phytosterols* (plant steroids).

Sesame oil may also protect against stomach cancer. In Korea, where stomach cancer is the most prevalent form of malignancy, researchers at the Seoul National University College of Medicine have discovered that Koreans who frequently consume sesame oil have a decreased risk of stomach cancer. Other foods linked to a reduced risk included mung bean pancake, tofu, cabbage, and spinach.

There's more to the anticancer story: When exposed to sesame oil in test tubes, human colon cancer cells stopped growing and multiplying, according to a study conducted at Maharishi International University in Fairfield, Iowa. Other vegetable oils had a similar effect, although it was not as dramatic as that of sesame oil. The researchers pointed out that vegetable oils, including sesame oil, appear to have anticancer properties that warrant further investigation.

Applied topically, sesame oil is a therapeutic skin treatment widely used in Ayurveda (pronounced eye-yuhr-VAY-dah), a system of medicine used for more than 5,000 years in India. Ayurveda treats diseases holistically, with nutrition, exercise, meditation, and other lifestyle-management approaches. In another study at Maharishi International University, researchers observed that sesame oil, as well as safflower oil, inhibited the growth of malignant melanoma cells in lab dishes. Malignant melanoma is the deadliest form of skin cancer.

Sesame oil is used mostly in cooking and is a popular oil for preparing Asian dishes. A note of caution: Sesame seeds and sesame oil contain allergens, which can provoke an allergic reaction in susceptible people.

Wheat Germ and Wheat Germ Oil

At the heart of a kernel of wheat grain is the germ, or the sprouting portion of the kernel. It's an excellent source of vitamin E, protein, B-complex vitamins, and various minerals, including calcium, magnesium, phosphorus, copper, and manganese. Wheat germ is usually removed from the grain during milling because its fat content can turn the flour rancid. Wheat germ oil is extracted from wheat germ and sold as a nutritional supplement. Fifty-eight percent of the essential fatty acids in the oil are omega-6 fats. Wheat germ oil is also high in vitamin E, an important antioxidant.

From wheat germ oil comes the supplement octacosanol, an alcohol derivative of the oil. Athletes have long used octacosanol in the belief that it will improve energy, strength, and reaction time, although no research has turned up any proof that it directly improves endurance or physical performance. Some studies suggest that octacosanol may help regulate cholesterol and prevent platelet stickiness (which can lead to abnormal blood clotting).

There are no known adverse reactions or side effects to using wheat germ, wheat germ oil, or octacosanol. In fact, wheat germ is an excellent, nutrient-packed food to include in the diet, although it is rather high in calories (360 calories in a 3½-ounce serving).

Wheat germ oil is available in capsules and in liquid form. Follow the manufacturer's suggest dosage if you decide to supplement with wheat germ oil.

Hempseed Oil

You may have heard or read about a nutritional oil called hempseed oil. It is extracted from the seeds of the hemp plant, also the source of the narcotic marijuana. But you won't get high from hempseed oil. Recently, newer hemp cultivars have been developed that yield a seed that is practically devoid of THC (tetrahydrocannabinol), the narcotic in hemp.

Still, it is illegal to grow the hemp plant in the United States, so hempseed oil is manufactured mostly in Canada, the United Kingdom, and parts of Asia and imported for use in cosmetics and toiletries. Hempseed oil is also available as a dietary supplement from various companies that market nutritional oils.

Hempseed oil is loaded with essential fats: omega-6 fats (58 percent), omega-3 fats (20 percent), monounsaturated fats (11 percent), and GLA (1 to 3 percent).

There have been very few studies conducted on the health benefits of hempseed oil, however, so it is not yet known whether it can be used effectively to treat diseases that benefit from essential fatty acid supplements.

A red flag: If you consume hempseed oil, you may test positive in a urine drug test for metabolites called cannabinoids, which are also evident in urine after marijuana use.

Part III

More Good-for-You Fats

SEVEN

Olive Oil: The Master Monounsaturated Fat

You're lunching on a Greek salad, built on a bed of green, leafy lettuce piled high with tomatoes, olives, and crumbled feta cheese. The salad is drizzled with a dressing made from extra-virgin olive oil with some vinegar and spices thrown in for added zing. With each forkful, you're plying your body with oleic acid, squalene, oleuropein, hydroxytyrosol, tyrosol, and other substances with hard-to-pronounce names.

Sound unappetizing? Don't let these names scare you. They designate an array of healthy food components, all with some amazing disease-fighting properties. You happen to be lunching on a combination of foods that has been eaten for hundreds of years in the Mediterranean, where rates of heart disease and cancer are among the lowest on the planet.

The so-called Mediterranean diet refers to the cuisine eaten primarily in Greece and southern Italy. It is characterized by low consumption of red meat; lots of fruits, vegetables, and grains; moderate amounts of red wine; and loads of olive oil. In fact, the fats derived from olive oil make up more than a third of the total daily calories for people from these countries. In Greece alone, the diet provides up to 42 percent of its calories as fat, primarily from olives and olive oil. Despite their high-fat eating habits, Greeks living in Greece enjoy the longest life expectancies of any group of people in the world.

But so healthy is the Mediterranean diet that from decades of research, it appears that if you eat like this, you're least likely to die of anything (with the exception of very old age). And olive oil is a prime reason.

What's so special about olive oil?

Rich in monounsaturated fats, olive oil has been used for centuries to maintain the suppleness of skin, give sheen to hair, aid in digestion, relieve muscle pain, and lower blood pressure. Ancient Egyptians used olive oil for preserving mummies, and the Bible mentions the oil frequently as a medicine for healing wounds and anointing the sick.

Among unsaturated fats, olive oil stands alone as being resistant to oxidation, a possible factor in heart disease. Because of this attribute, olive oil has been described as the safest fat to eat because it doesn't readily oxidize and give off health-damaging free radicals. More recently, olive oil has been shown to conserve heart-protective HDL cholesterol. Further, olive oil also seems to guard Mediterranean people from certain forms of cancer, specifically cancers of the colon, breast, endometrium, and prostate. Another plus for olive oil: Antioxidant vitamin E does a better job of fighting off free radicals if olive oil is present in your diet.

The secret to olive oil's power lies in its host of healthy constituents mentioned earlier. Let's return to those funny-sounding names for a moment and delve into what they do so you can understand the powerful effect they can have on your health.

Oleic Acid: A Disease-Stopping Fatty Acid

The chief monounsaturated fatty acid in olives and olive oil is an omega-9 fatty acid called *oleic acid*, which constitutes 80 percent of the oil. Oleic acid is also distributed in canola (see Table 8 in this chapter), almond, avocado, canola, peanut, safflower, and sunflower oils. It is what gives all of these oils, including olive oil, their ability to withstand higher cooking temperatures. So healthy is oleic acid that other vegetable oils have been specially modified to pump up their content of this fatty acid. Examples include high-oleic sunflower oil and high-oleic soybean oil. There is also an appreciable amount of oleic acid in meat, but the huge amounts of saturated fat in meat cancel out any good health deeds its oleic acid might do.

The high percentage of oleic acid in olive oil protects your cardiovascular health and may decrease your risk of heart attack and stroke, according to a growing body of research. Specifically, oleic acid in olive oil:

- Influences blood levels of harmful LDL cholesterol (the lower your intake of oleic acid, the higher your LDL cholesterol levels)

- Increases levels of protective HDL cholesterol

- Blocks the abnormal buildup of fat and plaque in arteries, which can lead to atherosclerosis, the narrowing and thickening of artery walls

- Lowers triglycerides, a blood fat that in excessive amounts can contribute to heart disease

Olive oil clearly does your heart good. So if you're interested in preventing and treating heart disease, include olive oil as one of the main good fats in your diet.

Squalene: Secret Cancer-Fighting Ingredient

Olive oil is well endowed with a vitaminlike chemical called *squalene*. Squalene is also present in shark liver oil and human sebum, an oily secretion that lubricates your skin.

Some rather fascinating statistics: The average intake of squalene in the United States is 30 milligrams a day, whereas in Mediterranean countries, the typical intake is 200 to 400 milligrams a day. Because of this dietary disparity, many scientists feel that the low incidence of cancer in Mediterranean countries is somehow linked to the higher intake of squalene obtained through olive oil.

Indeed, experiments with squalene show that it inhibits colon, lung, and skin tumors in rodents. Squalene is believed to work by suppressing the action of a key enzyme required for the growth of cancer cells. To date, no human studies have been conducted on the anticancer effect of squalene, although some scientists are convinced that this chemical may turn out to be the true cancer fighter in olive oil.

But with regard to colon cancer, a team of researchers at the Institute of Health Sciences in Britain compared colon cancer rates, diets, and olive oil consumption in twenty-eight countries, including Europe, Britain, the

United States, Brazil, Colombia, Canada, and China. They discovered that countries with a diet high in meat and low in vegetables had the highest rates of colon cancer. But olive oil was associated with a lower risk, leading the researchers to conclude that olive oil may have a protective effect on the development of colon cancer. The study didn't single out squalene as the reason, but as previously noted, other scientists have emphasized that it could be responsible for the cancer-fighting effect of olive oil.

Among squalene's other functions in the body are to act as an intermediate, or a go-between, in the synthesis of cholesterol. In essence, it helps regulate how much cholesterol is made and may even inhibit the abnormal production of this fatty substance. In a study of elderly patients, the combination of squalene (860 milligrams) with a cholesterol-lowering drug called pravastatin (10 milligrams) did a better job of lowering harmful LDL cholesterol and increasing beneficial HDL cholesterol than treatment with the drug by itself.

Oleuropein: Special Antioxidant Powers

Technically, oleuropein is a polyphenol, a term that describes its chemical structure. Distributed widely in plant foods, polyphenols are disease-fighting substances that boast a lengthy resume of health-promoting accomplishments, particularly in cardiovascular health and cancer protection. They are also responsible for the characteristic stability of olive oil, preventing it from turning rancid.

Oleuropein is the most abundant polyphenol in olives and olive oil. It gives olives their bitter taste, but more important, it acts as a mighty antioxidant to snuff out harmful free radicals. Experiments show that oleuropein can squelch the superoxide radical, a nasty molecule that inflicts tissue injury by damaging structural lipids in cell membranes. What's more, oleuropein can shut down a dangerous free-radical–generating process in the body.

There is mounting interest in oleuropein as a cholesterol fighter. In lab dishes, it has demonstrated potent antioxidant activity against the oxidation of LDL cholesterol—a process initiated by free radicals. When LDL is oxidized by free radicals, white blood cells in artery linings start attracting excessive amounts of LDL. The oxidized LDL forms lesions on the inner arterial walls, and cholesterol is deposited in these lesions—a process that leads to atherosclerosis.

There's more: Oleuropein appears to keep body-invading bacteria at bay, say investigators at the University of Messina in Italy. They found that oleuropein curbed the growth of various bacterial strains, specifically those that attack the intestines and urinary tract.

Hydroxytyrosol: A Free-Radical Attacker

Another powerful antioxidant in olives and olive oil is a polyphenol called *hydroxytyrosol,* a biological "chip off the old block," formed when it splits from oleuropein during an enzyme-controlled reaction. In the body, hydroxytyrosol combines forces with oleuropein, creating a dynamic antioxidant duo that is more powerful against free radicals than even vitamin C, another potent antioxidant.

Separately, hydroxytyrosol is an effective antioxidant in its own right. An interesting study conducted with rats at the University of Milan found that hydroxytyrosol prevented oxidative stress caused by exposure to passive smoking. Oxidative stress is a harmful condition that occurs when free radicals outnumber antioxidants in the body. Quite possibly, eating olive oil may fortify your bodily defenses against the adverse health effects linked to secondhand smoke.

Like oleuropein, hydroxytyrosol prevents the dangerous oxidation of LDL cholesterol and suppresses abnormal blood clotting—two benefits that make this healthful chemical a worthy dietary defense against heart disease.

Also, a recent Italian study discovered that hydroxytyrosol stopped cultured leukemia cells from multiplying and caused them to "commit suicide." This is very preliminary research, but promising nonetheless.

Tyrosol: Heart Helper

The third major polyphenol in olives and olive oil is tyrosol. Like its companion polyphenols, tyrosol is an antioxidant that prevents the oxidation of LDL cholesterol and keeps free radicals from attacking cell membranes.

Tyrosol may have anticancer properties as well. In one study, researchers in Italy discovered that tyrosol protected human colon cells in lab dishes from turning cancerous in response to oxidation.

Many scientists believe that tyrosol and the other polyphenols in olive oil have blood pressure–lowering properties. In one study, patients with high blood pressure who followed an olive oil–enriched diet were able to discontinue taking their blood pressure pills.

Why does tyrosol have such a profound benefit? Researchers chalk it up to the antioxidant effect of polyphenols against free radicals. The production of free radicals blocks the production of nitric oxide, a blood vessel relaxant that helps keep blood pressure down. Thus, eating more polyphenols—including those from olive oil—may help keep your blood pressure in check.

Maximize the Health Benefits of Olive Oil

Here's something else you should know: You can get even more health-protecting power from olive oil by adding it to tomato-based recipes. That's because combining olive oil with tomato products soups up the protective powers of lycopene, an antioxidant that may help prevent heart disease and cancer. In a study comparing the addition of olive oil versus sunflower oil to tomato products, only the mixture of olive oil with tomato increased antioxidant activity.

Purchasing Olive Oil

There is a dizzying array of various types of olive oil on supermarket shelves: extra-virgin olive oil, fine virgin olive oil, virgin olive oil, light olive oil, and so on. Not only that, olive oils are manufactured in more than twenty-two different countries, with widely different soil conditions, and are extracted from a large variety of olives, each with its own characteristics. All of these factors account for huge variations in the fatty acid and nutrient content of olive oil.

That being the case: Which olive oil is best, and does it even matter?

Some olive oils are healthier and more potent than others, so it does matter which type you buy. Basically, olive oils differ in their concentration of fatty acids and antioxidants—a difference largely due to the oil extraction process. The less heat and fewer chemicals used to extract the oil, the more nutritious the olive oil is.

Olive oil contains 115 calories per tablespoon—approximately the same number found in other cooking or salad oils. You may be able to save calories, however, because olive oil has a greater flavor and aroma than other oils, and that means you can use less when cooking. Olive oil is also cholesterol-free.

Table 7

PROPERTIES AND USES OF OLIVE OIL

Type	Processing Method	Color/Taste	Uses	Healthful Properties
Extra-virgin	Obtained from the first pressing of the ripe fruit by mechanical and physical methods. No heat or chemicals are used. The only vegetable oil that is not chemically processed.	Pale yellow to bright green; fruity taste; wide range of flavors and aromas.	Salad dressings; dipping bread.	Contains all the fatty acids and antioxidants found in the olives from which it is made. High levels of squalene. The only edible oil that contains polyphenols. Appears to be more protective against LDL oxidation than other olive oils.
Virgin	A second-press oil; however, some are processed using chemicals.	Colorless; good flavor.	Salad dressings; dipping bread; cooking. (Not widely available in the United States.)	Contains a negligible amount of polyphenols and squalene.
Fine	A blend of extra-virgin and virgin olive oils.	Varies in level of fruitiness.	Salad dressings; dipping bread; cooking.	Combines some of the nutrient characteristics of extra-virgin and virgin olive oils.

Table 7 (continued)

PROPERTIES AND USES OF OLIVE OIL

Type	Processing Method	Color/Taste	Uses	Healthful Properties
Pure olive oil	Produced by chemically extracting oil from the remaining pulp left over after cold-pressing.	Yellow to green; tasteless.	A general-purpose olive oil.	Contains little or no healthful substances.
Light	This is pure olive oil blended with a tiny amount of extra-virgin olive oil.	Lightest in color of all olive oils; generally tasteless. ("Light" or "lite" refers to color, not caloric content. All olive oils contain roughly the same amount of calories.)	Recommended for cooking.	Contains the same amount of beneficial mono-unsaturates as other olive oils, but contains few other healthful substances.

Table 7 provides a comparison of the various types of olive oil available. Clearly, your best bet is extra-virgin olive oil. But even then, there are differences in levels of two important components: oleic acid and squalene.

Some cancer researchers feel that the ratio of oleic acid to linoleic acid, an omega-6 fatty acid, is the best indicator of an olive oil's cancer-fighting power. A higher proportion of oleic acid to linoleic acid is considered desirable. As noted, other researchers feel that the squalene in olive oil is responsible for its anticancer benefit.

In 1999, *Prevention Magazine,* a leading health publication, tested various extra-virgin olive oils for their ratio of oleic acid to linoleic acid and for their squalene content. In an article published in the September 1999 issue, writer Mike McGrath reported that olive oils from Spain, Italy, and Greece

Table 8

THE OTHER MIGHTY MONOUNSATURATE

Another good fat with plenty of virtues is canola oil. Among cooking oils, it contains the lowest level (6 percent) of artery-clogging saturated fat and boasts 61 percent of heart-healthy oleic acid, making it second only to olive oil in oleic acid content. There are also considerable amounts of linoleic acid (22 percent), an omega-6 fatty acid, and alpha-linolenic acid (11 percent), an omega-3 fatty acid, in canola oil.

Canola oil is pressed from the seeds of the rapeseed plant, a relative of the mustard plant that has been cultivated in China and India for more than 4,000 years. Originally, oil from the rapeseed plant was available as a food oil solely in Europe and Canada, but not in the United States because it was accused of causing heart abnormalities in rats and consigned to industrial uses only. A new cultivar of the plant was bred to be very low in a natural but potentially toxic fatty acid called erudic acid, and in 1985, its oil was approved by the FDA for cooking. The FDA considers the low erudic acid content (0.6) of canola oil safe for human consumption.

At first, the oil was required to be labeled "low-erudic-acid rapeseed oil," but in 1988, the FDA permitted the product to be called canola oil, the name used in Canada, where most of the oil is produced. In the United States, some canola oil is made in Indiana, Kentucky, and Tennessee.

Like olive oil and many other good fats, canola oil has been found in research to lower total cholesterol and LDL cholesterol and normalize the clotting activity of platelets, thereby reducing the risk of heart disease. In animal studies, it has been shown to protect against dangerous arrhythmias (irregular heartbeats).

Canola oil contains 120 calories per tablespoon, is available as cooking oil, and is found in margarines.

---

scored the highest for oleic acid ratios and squalene content. As for squalene, one oil stood out for having the highest amount of this protective nutrient: St. Helene Extra Virgin, an olive oil made in California.

Best advice: Buy extra-virgin olive oil, preferably from Mediterranean countries, or check out the California-made oil mentioned earlier to reap the health-protective benefits of this exceptionally good fat.

See Table 8 for a discussion of another healthy monounsaturated fat: canola oil.

The New Heart-Saving Fats

Attention spread lovers: Want to have your margarine and butter—and eat it, too? Then check out the margarine spreads Benecol and Take Control, which are designed specifically to reduce cholesterol levels in the blood.

In 1999, the U.S. Food and Drug Administration (FDA) approved both products with the designation *functional foods*. Technically, the term *functional food* refers to a food product that is beneficial to health. In its position paper on functional foods, the American Dietetic Association formally defines such foods as "any modified food or food ingredient that may provide a health benefit beyond the traditional nutrients it contains." Benecol and Take Control fit this definition because they are formulated with ingredients designed to improve cardiovascular health.

Both margarines are predominantly monounsaturated fats (mainly canola oil) fortified with plant sterols, including *sterol esters* and *stanol esters*. Similar in structure to cholesterol found in animals, plant sterols are present in small quantities in many fruits, vegetables, legumes, nuts, seeds, grains, and other plant sources. Benecol contains plant stanol esters from a compound found in pinewood pulp, and Take Control uses plant sterol esters extracted from soybean oil. If your cholesterol is moderately high,

Benecol and Take Control may help you control it, reducing your risk of artery damage.

How Cholesterol-Lowering Margarines Work

Under normal conditions, LDL cholesterol delivers cholesterol to cells to make cell membranes, sex hormones, and other substances. But in excessive amounts, LDL cholesterol—the so-called bad cholesterol—can lead to artery damage. By contrast, HDL cholesterol—known as the good cholesterol— picks up cholesterol from cells and shuttles it to the liver, where the cholesterol is turned into bile acids (*biliary cholesterol*) and secreted in the intestines.

Essentially, plant sterols make it harder for your intestines to absorb cholesterol from the small intestine. Plus, they increase the excretion of both biliary cholesterol and dietary cholesterol.

Specifically, plant sterols work in the following manner: When your body soaks up cholesterol from the small intestine, the cholesterol is first wrapped in a *micelle*, a tiny droplet made of lipids and emulsifiers from bile. Because plant stanol/sterol esters are very similar in shape to cholesterol, micelles mistake them for cholesterol. If a plant stanol/sterol ester gets into the micelle first, there's no room for cholesterol. In essence, plant sterols and cholesterol compete for entry into the micelles. Unable to enter the micelle, cholesterol is blocked from absorption into the bloodstream. The net effect of displacing cholesterol is reduced levels of LDL cholesterol in the blood and the preservation of levels of HDL cholesterol. Cholesterol-lowering margarines do not raise HDL cholesterol, block saturated fat, or lower triglycerides, however.

A Look at the Scientific Proof

Have these products been scientifically proven to reduce cholesterol? Yes. Here's the scoop:

- A yearlong study conducted at the University of Helsinki in Finland compared patients using Benecol with patients using regular margarine (the control group). Benecol (roughly 2 tablespoons eaten daily for one

year) reduced overall cholesterol levels by 10 percent and LDL cholesterol by 14 percent in people whose cholesterol was mildly elevated (between 200 mg/dl and 240 mg/dl). What this means: Let's say your total
cholesterol is 230; eating Benecol could lower it by as much as 23 points.

As for Take Control, research shows that it cuts LDL cholesterol by
10 percent, when patients used one or two pats a day, and by 17 percent
when used as part of a heart-healthy diet.

● For more dramatic results, try combining cholesterol-lowering spreads
with cholesterol-lowering drugs called *statins* (but check with your
physician first). Statins lower LDL cholesterol by curtailing your body's
production of cholesterol.

In twenty-two women diagnosed with heart disease, Benecol alone
dropped total cholesterol by 13 percent and LDL cholesterol by 20 percent. Combined with a statin drug, Benecol cut cholesterol levels even
further: total cholesterol by an extra 11 percent and LDL cholesterol by
an extra 16 percent in just twelve weeks.

In a third of the women tested, Benecol alone normalized cholesterol.
In others, the combination of Benecol and the statin drug was found to be
more effective than the statin alone. What's more, the combo allowed
some patients to take a lower dosage of medication, and in some cases,
eliminated the need for the drug altogether. This is certainly good news,
since statins often have side effects, including liver problems, stomach
upset, and muscle pain.

These findings are monumentally significant for women. Here's
why: If you're like most women, you probably think heart disease is a
"guy thing"—you know, like football, hunting, or John Wayne movies.
But that perception is not only false, it's dangerous. Among women,
heart disease claims more lives than all forms of cancer combined, even
breast cancer. Statistics show that heart disease strikes 36 percent of
women, while breast cancer affects 4 percent. One in two women will
die from a first heart attack, compared to one in four men. And among
those who do survive, 20 percent will have a second heart attack within
four years, compared to 16 percent of men. The study cited earlier simply means that now women have one more weapon—plant stanol/sterol
esters—to get their cholesterol levels under control and defend themselves against heart disease.

So that men don't feel left out: Benecol (5.1 grams a day) reduced to-tal cholesterol by 12 percent and LDL cholesterol by 17 percent in men (and women) undergoing statin therapy but who still had elevated LDL cholesterol, according to an eight-week study conducted at the Cooper Institute in Dallas, Texas.

- Plant sterols complement the cholesterol-lowering effects of a low-fat diet, which is recommended as a way to reduce cholesterol. Studies show that people who strictly adhere to low-fat eating, can slash cholesterol by 15 to 37 percent. Often, though, it's tough to stick to a low-fat diet be-cause fat—especially the saturated variety—is so hard to give up.

  If that sounds all too familiar, swap your favorite fats such as mar-garine and butter for a product such as Benecol or Take Control. In a study conducted at the University of Kuopio in Finland, people who ate wood- or vegetable-derived plant sterols while following a low-fat diet brought their total cholesterol and LDL cholesterol down even lower than by following a low-fat diet alone.

  What constituted a low-fat diet in this study? The experimental diet provided 28 to 30 percent of total calories from fat (8 to 10 percent as saturated fat, 12 percent as monounsaturated fat, and 8 percent as polyunsaturated fat), 20 percent of calories from protein, and 50 to 52 percent from carbohydrates.

- Benecol and Take Control work equally well. A study of ninety-five pa-tients conducted in the Netherlands compared the effects of margarines fortified with either plant stanol esters or plant sterol esters. Both types of margarines brought total cholesterol and LDL cholesterol down by as much as 8 to 13 percent.

- Plant sterols may turn out to be effective dietary tools for heading heart disease off at the pass, even reversing its course. Based on the expected LDL cholesterol reduction of 10 to 14 percent, scientists estimate that plant sterols may be able to cut the risk of heart disease by a whopping 25 percent. Already, in experiments with mice, plant sterols have inhib-ited the formation of lesions inside artery walls, reversing the develop-ment of atherosclerosis. Of course, we need more research, particularly with humans, to see how this all shakes out, but plant sterols certainly show enormous potential in preventing heart disease.

Using Cholesterol-Lowering Margarines

Benecol and Take Control most definitely count as healthy fats. The fact that they are made from monounsaturated fats is a plus because mono fats are practically immune to oxidation, a factor in atherosclerosis.

It should be noted that Benecol contains minute amounts of trans-fatty acids—less than 0.5 grams per 8-gram serving. (Foods totaling less than 0.5 grams can claim to be transfree.) Take Control contains 0 grams of trans-fatty acids per serving.

Here are some additional tips for using these products to your best advantage:

- With Benecol, the manufacturer (McNeil Consumer Healthcare) recommends three daily servings (½ teaspoons each) to move your cholesterol into a healthier range. Two servings a day (2 tablespoons) of Take Control are recommended by its manufacturer (Lipton).

- You can use these products at a single meal or divide your servings over three meals. A recent study demonstrated that a daily serving of 2.5 grams of plant stanol esters eaten once at lunch lowered LDL cholesterol just as effectively as spreading the serving over three meals (0.42 grams at breakfast, 0.84 grams at lunch, and 1.25 grams at dinner). These findings suggest that plant sterols stay in the intestinal tract for several hours after meals.

- These spreads should always be used in place of—not in addition to—butter or margarine. You can spread them on toast, vegetables, crackers, potatoes, bagels, or other favorite foods.

- Try cooking, baking, or frying with Benecol Regular Spread. Take Control is not recommended for cooking.

- Benecol and Take Control come in tubs, like other margarines in use today, making it easy to measure out portions.

- Figure these margarines into your daily calorie count if you're watching your weight. Benecol Regular contains 45 calories per serving; Benecol Light, 30 calories per serving. Take Control contains 50 calories per serving; Take Control Light, 40 calories per serving.

- Use these products as part of an overall cholesterol-reduction program that includes eating less saturated fat and trans-fatty acids, exercising,

Table 9

WHAT ARE FAKE FATS?

Plant stanol/sterol esters are not the same substances as *fat replacers.* Fat replacers are additives concocted from carbohydrates, protein, and other fats to trick your mouth into believing you're getting a high-fat treat, but without the calories. They resemble fats in all ways except two: They don't plug arteries, and they're low in calories.

Fat replacers and fat substitutes are categorized as follows:

• **Carbohydrate-based fats.** Formulated from starches and fibers, carbohydrate-based fats are used in frozen desserts, puddings, cake frostings, margarines, and salad dressings to replace the real fat. Examples include polydextrose, a partially absorbable starch that supplies about 1 calorie per gram (versus 9 calories per gram from fat), and maltodextrin, a starch made from corn. Cellulose and gum are two types of fibers used to manufacture fat replacers. When ground into tiny particles, cellulose forms a consistency that feels like fat when eaten. Cellulose replaces some or all of the fat in certain dairy-type products, sauces, frozen desserts, and salad dressings. Gums such as xanthan gum, guar gum, pectin, and carrageenan are used to thicken foods and give them a creamy texture. Added to salad dressings, desserts, and processed meats, gums cut the fat content considerably. Side effects of carbohydrate-based fat replacers include gas, bloating, and cramping.

• **Protein-based fat replacers.** These are formulated from milk or eggs, created by heating and blending milk or egg white proteins into mistlike particles—technically known as *microparticulated protein.* Like real fat, these processed proteins feel creamy on the tongue. You'll want to avoid this type of fat replacer if you are allergic to milk or eggs.

• **Fat-based fat replacers.** When serving a fattening dessert, have you ever jokingly told your guests that "all the calories have been taken out"? It can be done! Food technologists can chemically change the properties of fats to remove or drastically cut fat calories. The result is fat-based fat replacers and fat substitutes. Some pass through the body unabsorbed, making them calorie-free. Others can even be used in cooking and frying, unlike carbohydrate- and protein-based fat replacers.

One of these is Olestra from Procter & Gamble. Technically, Olestra is a *sucrose polyester,* meaning a combination of sugar and fatty acids. Your body can't digest Olestra, so it's classified as a noncaloric product.

Table 9 (continued)

WHAT ARE FAKE FATS?

According to the manufacturer, Olestra can be substituted for fats and oils in foods without loss of flavor. It has the same cooking properties as fats and oils and can be used in shortenings, oils, margarines, ice creams, desserts, and snacks. Also, Olestra may block the absorption of vitamins A, D, E, and K. Side effects include cramping and loose stools.

Other fat-based fat replacers are emulsifiers, which reduce fat and calories by replacing the shortening in cake mixes, cookies, icings, and other products, and caprenin, a cocoa butter–like ingredient used in candy.

• **Combinations.** Some fat replacers are formulated from a combination of carbohydrates, proteins, and fats. Combination-type fat replacers are found in ice cream, salad oils, mayonnaise, spreads, sauces, and bakery products.

The key to enjoying any new fat-free product is to include it in your diet using a tried-and-true nutrition maxim—with moderation. Remember, too, that even though a food may be fat-free, it's not necessarily calorie-free. If you eat a lot of fat-free products, you could be overconsuming calories and packing on pounds as a result.

quitting smoking, losing weight, and taking cholesterol-lowering drugs, if prescribed. Benecol is also available in snack bars and salad dressings.

● Eating more than the recommended amount will not lower cholesterol more than the 10 to 14 percent observed in studies.

● While these products are effective for fighting elevated cholesterol, they can also be used by anyone who wants to guard against too-high cholesterol.

● Cholesterol-lowering spreads have been intensely evaluated for safety and have passed with flying colors, earning the Generally Recognized As Safe (GRAS) designation from the FDA.

Part IV

The Fat-Burning Fats

Conjugated Linoleic Acid: Health Insurance in a Pill

Talk about confusing: Some fats have been linked to the most life-threatening diseases on the face of the earth—heart disease, cancer, diabetes, and obesity. Even so, there is yet another fat that has the power to prevent and treat the very same diseases!

How can this be?

Meet conjugated linoleic acid (CLA), another body-friendly fat that provides excellent health insurance against a variety of ills. Discovered in 1978 at the University of Wisconsin, CLA is a naturally occurring fatty acid present in dairy products (most notably, milk fat), as well as in meat, sunflower oil, and safflower oil. It is formed when the bacteria in a cow's gut breaks down the essential fatty acid, linoleic acid, in the food the animal eats.

In 1996, CLA became available as a diet product derived from sunflower oil. Ads for CLA note that the nutrient may be missing from many diets (presumably since some people tend to eat less meat and high-fat dairy products). Since its discovery, there has been an explosion of research on CLA—all pointing to some rather amazing results, particularly in the area of weight loss and weight management.

Here's a rundown on the health-protecting properties of this rather remarkable supplement. CLA:

Melts Pounds

At any given time, 33 to 40 percent of women and 20 to 24 percent of men are trying to knock off pounds, says the American Society of Bariatric Physicians (physicians who specialize in weight loss). Wouldn't it be great if we could pop a pill to melt fat, but experience no untoward side effects? CLA may very well be that pill.

Much research with animals documents that CLA is a potent fat-fighter. In tests with mice, CLA pared down body fat (even at night), curbed appetite, boosted metabolism—and spot-reduced the animals' abdominal area!

Although the mechanism of CLA's fat-fighting action is unclear, researchers have found that it encourages the breakdown of fat, stifles a fat storage enzyme called *lipoprotein lipase,* and kills off fat cells. All three factors could be responsible for CLA's fat-burning benefits—at least in animals.

Some researchers believe that CLA may interact with *cytokines,* proteins that are involved in energy production and fat metabolism. They theorize that CLA somehow causes dietary protein, carbohydrates, and fats to be used by cells for energy and muscle tissue growth, rather than be stored as fat.

Another observation: CLA investigators say that the supplement does not shrink fat cells (like weight-loss diets do), but rather, it keeps them from enlarging. Enlarged fat cells are the main reason we get pudgy.

But does CLA work as well in humans? A string of new evidence from human trials shows that while CLA will not decrease your body weight, it does something better: It reduces body fat and builds body-firming muscle. Muscle is metabolically active, meaning that it burns fat even while you're resting. Also, the more muscle you have, the faster your metabolism. A fast metabolism translates into better weight loss and better weight control.

In one study, conducted at the University of Wisconsin in Madison, volunteers took 3 grams of CLA for six months. The findings showed that CLA kept people from gaining weight, and also increased their muscle tissue. That's good news for people entering middle age. Here's why: We tend to get fatter and lose muscle with age; thus, supplementing with CLA just might prevent middle-age spread.

In a study at Kent State University twenty-four men, ages nineteen to twenty-eight, supplemented with either 7.2 grams a day of CLA, or a vegetable oil placebo, while participating in a six-week bodybuilding program.

The CLA-supplementers experienced gains in the following areas: greater arm girth, overall body mass, and improved leg strength. The researchers commented that "apparently CLA acts as a mild anabolic agent."

In another human study, twenty nonobese people (ten men and ten women) were given just over a gram of CLA or a placebo with breakfast, lunch, and dinner. They were instructed not to change their diet or exercise habits.

At the end of three months, the researchers measured both the weight and body fat percentage of the study participants. Even though there was not much difference in weight loss between supplementers and nonsupplementers, there was a huge difference in body fat percentage. The CLA-supplementers dropped from 21.3 percent (average body fat) to an average of 17 percent. While it might not sound like much, a reduction of a few points in body fat percentage can make a huge difference in how lean and firm you look. The people taking CLA lost mostly body fat—the ideal situation in any trim-down program.

If you're already overweight, CLA banishes pounds, too, according to a recent study published in the *Journal of Nutrition*. Sixty overweight people randomly took either a placebo or CLA for twelve weeks. The CLA dosage ranged from 1.7 grams to 6.8 grams daily. By the end of the experimental period, those who supplemented with 3.4 grams of CLA daily had dissolved their body fat by six pounds, on average. The researchers concluded that supplementing with 3.4 grams a day is enough to pare down pounds, plus manage your weight.

Another study found that dieters who stopped dieting, but continued to take CLA, were more likely to gain body-firming muscle afterward, rather than ugly fat pounds. This finding hints that CLA may be a great supplement for postdieting maintenance.

One thing to keep in mind: Don't watch for lost pounds on your scale if you're taking CLA. The research cited above clearly points to the fact that CLA gets rid of fat but enhances muscle. Because muscle weighs more than fat, the best way to see the effects of CLA is by checking the notches on your belt, rather than by weighing yourself on a scale. Scales simply don't give the best picture of whether you're tubby or toned, especially if you're supplementing with CLA.

CLA is widely promoted as a fat-burning, energy-boosting agent and is included as a primary ingredient in many weight-loss supplements now on health-food-store shelves. CLA does not require a prescription.

Fights Cardiovascular Disease

Cardiovascular disease is among the most dreaded of all health problems—and for good reason. It is the number one killer of adults in the United States.

Now some astonishingly good news: In addition to diet, exercise, and other healthy lifestyle practices, one way to escape this killer may be supplementation with CLA. That's because, in experimental animals, CLA thwarts the formation of plaque, the fatty deposits in the lining of the artery walls that lead to atherosclerosis.

What's more, animals fed CLA have significantly reduced levels of low-density cholesterol, very-low-density cholesterol, and triglycerides (three nasty conspirators in heart disease).

Further, in an experiment with human blood, CLA blocked a potentially dangerous process called *platelet aggregation,* in which tiny clotting elements in the blood called platelets tend to stick together when they're not supposed to. Clots are apt to form, contributing to heart attack and stroke. CLA may thus help protect against heart disease, but further studies are needed to confirm this benefit.

Protects Against Cancer

Another of the most dreaded diseases known to man is also the second leading cause of death in the United States: cancer. More than 1,500 Americans die of cancer each day, according to the American Cancer Society. Mounds of research with animals reveal that CLA protects against breast cancer, discourages the growth of skin tumors, and shrinks cancerous prostate tumors.

CLA's cancer-fighting properties were discovered in a rather serendipitous way. While investigating carcinogens that occur in grilled meats, University of Wisconsin researchers found that CLA blocked the formation of cancer-causing substances. This amazing finding led to more intensive research on CLA's potential as a cancer-fighter. Many animal studies have since found that it suppresses mammary cancer and skin cancer.

Published in 1996, a large-scale study conducted by Finland's National Public Health Institute produced compelling evidence of CLA's anticancer benefit in humans. Women who drank milk regularly for twenty-five years slashed their odds of getting breast cancer by 50 percent, compared to nonmilk-drinking women. The investigators zeroed in on CLA as the likely agent for the protective effect, since the fatty acid is highly concentrated in milk fat.

A word of advice: There's no CLA in fat-free milk, so many medical experts are recommending that women switch to low-fat milk (which contains CLA) to get possible protection against breast cancer.

What's more, numerous studies have found that CLA is toxic to human breast cancer cells in lab dishes, and slows their growth. And, just-completed human studies in Finland and France indicate that CLA intake is associated with a reduced risk of breast cancer and its recurrence. Moreover, tests conducted with human cancer cells grown in cell cultures show that CLA also inhibits the growth of melanoma, colon cancer, and lung cancer.

Animal research reveals that CLA may also help prevent and reverse a wasting disease called *cachexia,* which occurs when the body burns up muscle to obtain energy for fighting diseases such as cancer. Cachexia compromises the quality of life and long-term survival of cancer patients.

The big question: Just how does CLA put up such a powerful shield against cancer?

Scientists believe that CLA may enhance immunity—in three possible ways. One is by acting as an antioxidant that fights disease-causing free radicals and thus fortifies cellular defenses. Second, CLA appears to reduce the formation of a potentially harmful type of prostaglandin that suppresses T-cells, which identify and destroy cancer cells. This not-so-friendly prostaglandin also shuts down the production of interleukin-2, an immune agent required by T-cells for growth. In other words, CLA bolsters your immune system by making sure immune cells can do their job of killing off cancer cells, without inference from bad prostaglandins. A third theory holds that grazing cows extract anticancer compounds from the pasture vegetables they eat and transfer them to milk. There may be yet-to-be discovered mechanisms at work, too, since CLA is made up of different active compounds that may produce different effects.

Defeats Diabetes

Some 2,200 Americans are diagnosed every day with diabetes. Though treatable, diabetes is the seventh leading cause of death in the United States, and nearly 16 million people have it, according to statistics from the American Diabetes Association. Treatment is multifaceted and includes diet, exercise, oral medication, and insulin therapy (depending on what type of diabetes you have).

Emerging new evidence indicates that CLA may be an effective natural therapy for diabetes as well. One study involving prediabetic rats found that CLA improved glucose tolerance (the ability to transport blood glucose into cells for use by the body) and normalized too-high levels of glucose in the blood. The researchers noted that CLA might turn out to be an important therapy for the prevention and treatment of type II diabetes, the most common form of the disease.

In a clinical trial, people with poorly controlled diabetes took 3 grams of CLA daily for twelve weeks and were able to bring their blood sugar closer to normal range. In addition, they slashed their triglycerides levels in half. (High triglycerides are frequently diagnosed in people with diabetes and represent a risk for heart disease.)

Outmaneuvers Osteoporosis

Osteoporosis is the most common human bone disease, with 25 million sufferers in the United States alone. In women, bone loss can start as early as ages thirty to thirty-five at the rate of 0.5 to 1 percent a year. In the first three to five years following menopause, this can accelerate to 3 to 7 percent a year, making it possible to lose 9 to 35 percent of bone mass. By age seventy, some women may lose up to 70 percent of their bone mass.

One of the many factors that spells trouble for bones is prostaglandin E2 (PGE2), one of the so-called bad prostaglandins. Small amounts of PGE2 assist in bone formation, but too much PGE2 inhibits bone growth and has been linked to osteoporosis and arthritis. If you overindulge in omega-6 fats (found in vegetable oils, processed foods, and fast foods), your body steps up its production of PGE2. In studies with rats given CLA, the CLA increased the rate of bone formation while arresting the production of PGE2.

As for human studies, researchers at the University of Memphis investigated the effects of CLA supplementation during weight training on bone mineral (the calcium phosphate in bones) and on certain immune factors. For twenty-eight days, the subjects (twenty-three experienced male weight trainers) took 6 grams a day of CLA, or 9 grams a day of an olive oil placebo. By the end of the experimental period, bone mineral content had increased in the CLA-supplementers; their immune status also improved.

Other than the studies described in this chapter, there has not been much research into CLA and osteoporosis. But stay tuned: Current knowledge

Table 10

IS LECITHIN A FAT-BURNING FAT, TOO?

One of the oldest "fat-burning" fats around is lecithin, a phospholipid found in cells and nerve membranes; in egg yolk, soybeans, and corn; and as an essential constituent of animal and vegetable cells. Lecithin helps process cholesterol in the body. It is loaded with the B-vitamin choline, which prevents fat from building up in the liver and shuttles fat into cells to be burned for energy.

Being so high in choline, lecithin has been dubbed a fat burner and is available supplementally in capsules, granules, and liquid form. But there's no proof that it helps you burn fat. Technically, lecithin is a fat, and like a fat, it provides 9 calories per gram, or about 250 calories per ounce. In fact, research shows that a side effect of supplementing with high doses of lecithin is weight gain.

about CLA suggests that this supplement represents an avenue of bone research clearly worth pursuing.

Supplementing with CLA

From the looks of the research, supplementing with CLA may keep you from prematurely going to your grave. With that possibility in mind, how much should you take?

Because most of the research on CLA has been conducted with animals, the appropriate dosage range for people is unknown. Generally, though, the dosage of CLA supplements ranges from 2 to 4 grams daily. Health care practitioners recommend that you take your CLA supplements with food.

Since CLA is so prevalent in food, can you get by with eating dietary sources?

Not really. There's only about 0.14 grams in a 3-ounce hamburger, 0.04 in a glass of whole milk, and 0.03 grams in an ounce of cheese. So gobbling hunks of cheese or swilling gallons of whole milk is not a good way to obtain CLA from the diet, since high-fat foods eaten in excess contribute to heart disease and other serious illnesses.

Although CLA is found in beef, lamb, and dairy products, changes in cattle feed, along with the public's avoidance of fat-rich diets, may mean that

Americans are getting 65 percent less CLA than in the past, according to CLA experts. Vegetarians get even less.

Still, taking 3 grams a day is about twenty to forty times the amount you normally get from food if you eat animal products.

Postscript: There appear to be no reported side effects to supplementing with CLA, probably due to the limited research that has been conducted thus far in humans. However, one human trial found that three months of CLA supplementation (4.2 grams daily) increased lipid peroxidation, a harmful cascade of events in which free radicals attack the lipid-rich membranes of cells.

MCT Oil: The Dieter's Fat

It's a dieter's dream: a fat that melts fat.

Does such a fat exist? You bet. Meet medium-chain triglyceride (MCT) oil, a special type of dietary fat that was first formulated in the 1950s by the pharmaceutical industry for patients who had trouble digesting regular fats. Since then, MCT oil has been used therapeutically to treat fat malabsorption, cystic fibrosis, and obesity. In the last fifteen years, it has been employed effectively as a sports supplement and as a nutritional supplement that aids in weight loss and helps boost energy.

MCT oil is processed mainly from coconut oil but does not seem to have any of the adverse side effects associated with tropical oils, such as elevated LDL cholesterol. It occurs naturally in many of the foods we eat and is quite plentiful in breast milk.

MCT oil is no ordinary fat. Although it tastes much like regular salad oil, that's where the similarities end. At the molecular level, MCT oil is structured very differently from conventional oils and thus has some unique properties.

Conventional fats are made up of long carbon chains, with sixteen or more carbon atoms strung together. MCT oil, on the other hand, has a much shorter carbon chain of only six to twelve carbon atoms—hence its name, medium-chain triglyceride. Because of its shorter chain, MCT oil is metabolized much more quickly than fatty acids from regular oils or fats.

Also, unlike other fats, medium-chain fats are more water soluble. They can be absorbed more easily through the intestinal wall, requiring fewer enzymes or bile salts. In the intestines, MCT oil is broken down into fatty fragments, which combine with a water-soluble protein in the blood. From there, MCTs go right into the bloodstream and are transported to the liver, where, inside the cells, they are rapidly oxidized or burned up.

Basically, very little of this remarkable fat leaves the liver; therefore, MCTs rarely end up being stored as body fat—a scenario that does not occur with conventional fats. So by using MCT oil, you can have your fat and eat it, too, because it is burned immediately for energy and therefore is not stored as body fat like conventional fats and oils are.

Anyone who is interested in weight control or exercise performance should explore the merits of supplementing with MCT oil. Here is a closer look at what this supplement can do for you.

Stimulates Thermogenesis

MCT oil is burned up so quickly that its calories are turned into body heat—a process known as *thermogenesis,* which boosts your metabolism—your body's food-to-fuel process. The higher your metabolism, the more calories your body burns.

Several studies in animals and humans have tested MCT oil's thermogenic effect. An often-cited study that looked into this effect was published in the journal *Metabolism* in 1989. To determine whether MCT oil affected thermogenesis differently from regular fats, ten men (ages twenty-two to forty-four) were fed with liquid diets containing 40 percent of fat as either MCTs or regular fats. After five days on this diet, thermogenesis increased significantly in the men who consumed MCTs—but remained unchanged in the men who ate regular fats. The researchers concluded that MCT oil stimulated thermogenesis to a greater degree than an equivalent amount of conventional fat did. They also noted that MCT oil was less likely to be stored, in contrast to conventional fats.

Revs Up Metabolism

Have you ever felt like you're losing fat at a snail's pace, or not at all?

Perhaps your metabolism has slowed down and needs a nudge. This often happens in response to low-calorie dieting. In fact, research shows that your metabolic rate drops considerably during a period of caloric restriction. With a sluggish metabolism, your body can't burn calories efficiently, and it's difficult to pare pounds, even though you're on a diet.

Try adding MCT oil to your food to counter this effect. Researchers in Czechoslovakia discovered that when obese patients, who were following a low-calorie diet, supplemented with MCT oil (1 tablespoon daily), their metabolic rate stayed elevated. Thus, incorporating MCT oil into your weight-loss diet may help prevent those frustrating plateaus that dieters so often hit.

Another study corroborates this benefit. In an experiment involving seven healthy men at the University of Rochester, researchers tested whether a single meal of MCTs would increase the metabolic rate more than a long-chain fat meal would. The men ate test meals containing 48 grams of MCT oil or 45 grams of corn oil, given in random order on separate days.

Metabolic rate increased 12 percent over six hours after the men ate the MCT meals but increased only 4 percent after corn oil was consumed. What's more, concentrations of fats in plasma (the liquid portion of blood) were elevated 68 percent after the corn oil meal, but did not change after the MCT meal. These findings led the researchers to speculate that replacing conventional oils with MCTs over a long period of time might be beneficial in weight loss.

Burns Fat

Because MCT oil revs up your metabolism, you can potentially burn more fat. In a single-blind, randomized, crossover study, twenty healthy men ingested a single dose of either 71 grams of MCT oil or canola oil. Blood samples were taken prior to the dosing, then at one-hour intervals over five hours following supplementation. Triglycerides, or blood fats, actually decreased 15 percent as a result of taking MCT oil, whereas fats increased 47 percent with canola oil. These findings are quite remarkable: MCT oil burns up and reduces fatty substances in the body. The ramifications of this finding are important for anyone who wants to achieve and maintain a trim physique.

Burns Calories

The best way to do away with stored body fat is to consume fewer calories than your body needs to meet its physical and metabolic demands. So to lose fat, you must create a calorie deficit, either by eating less, exercising more, or both. Now, according to a recent scientific study, there's another way to get that deficit—without sweating or depriving yourself—but by supplementing with MCT oil.

When eight healthy men at the University of Geneva took just 1 to 2 tablespoons of MCT oil as part of their normal diet, their daily caloric expenditure increased by 5 percent! On average, they burned 113 extra calories a day.

What does this mean to you? Suppose, for example, you're eating roughly 2,000 calories a day, with part of those calories coming from MCT oil. You'd effortlessly spend an additional 100 calories a day—the equivalent of walking for thirty minutes at a moderate pace. You'd burn extra calories, without any extra effort!

Enhances Muscle

A healthy goal of any fat-loss program is to preserve as much muscle tissue as possible. The more muscle you have, the better your body can burn fat. This is because muscle is the body's most metabolically active tissue. It burns calories, even at rest. Fat tissue is not as active.

Unfortunately, far too many diets overrestrict calories. Severe caloric restriction forces the body to start cannibalizing its own precious muscle tissue (including heart tissue) for energy.

You can guard your muscle by following a calorie-adequate diet, sticking to an exercise program that includes strength training—and supplementing with MCT oil. In a study at Calgary University in Alberta, Canada, healthy adults were placed on a low-carbohydrate diet supplemented with MCT oil. The researchers measured the use of fat and protein in the subjects' bodies and found that there was an increase in body fat burned and a decrease in muscle protein used for energy. In other words, MCTs helped burn fat and, at the same time, preserved lean muscle by preventing its breakdown.

Increases Endurance

Exercise is an important way to encourage fat loss. The harder and longer you exercise, the more fat you can burn. But often it's tough to exercise all-out, and one reason is low energy. That's where MCT oil can help.

First, it provides twice the energy of protein and carbohydrate (8.3 calories per gram versus 4 calories per gram for carbohydrates and protein) and is absorbed into the bloodstream as rapidly as glucose, the cellular fuel made available from the breakdown of carbohydrates.

Second, MCT oil is preferentially used as fuel for energy, instead of being stored by the body. Medium-chain fatty acid fragments can diffuse into the cell very quickly, where they are burned immediately for energy—at the same time as glucose. The ability of MCTs to enter the cells in this manner has a glucose-sparing effect, meaning that glucose and its stored counterpart, muscle glycogen, last longer without being depleted. The longer glycogen reserves last, the more energy you have for activities and fat-burning exercise.

To boost your endurance during exercise, take MCT oil with a carbohydrate source such as a sports drink. At the University of Capetown Medical School in South Africa, researchers mixed 86 grams of MCT oil (nearly 3 tablespoons) with two liters of 10-percent glucose drink to see what effect it would have on the performance of six endurance-trained cyclists. The cyclists drank a beverage consisting of glucose alone, glucose plus MCT oil, or MCT oil alone. In the laboratory, they pedaled at moderate intensity for about two hours and then completed a higher-intensity time trial. They performed this cycling bout on three separate occasions so that each cyclist used each type of drink once. The cyclists sipped the drink every ten minutes. Performance improved the most when the cyclists supplemented with the MCT/glucose mixture. The researchers did some further biochemical tests on the cyclists and confirmed that the combination spared glycogen while making fat more accessible for fuel.

Using MCT Oil in a Fat-Loss Diet

If you want to jump-start your weight-loss efforts, replace some of the carbohydrates in your diet with MCT oil. Carbohydrates such as rice, cereal, pasta, breads, fruits, potatoes, and sweet potatoes offer a mother lode of vital nutrients. But too much carbohydrate in your diet can be fat-forming, particularly if you're not very active. A carbohydrate overload triggers a surge of the hormone insulin into your bloodstream. Insulin activates fat cell enzymes that move fat from the bloodstream into fat cells for storage, and this action spells extra weight for many of us.

When you reduce your intake of carbohydrates, you suppress the release of insulin. Low insulin stimulates the release of another hormone, glucagon. As glucagon goes to work, it signals the body to start burning fat for energy.

Reducing carbohydrate intake to lose weight has long been the basis of many diets. And for good reason—it works. But there are penalties. Carbohydrate is your body's leading nutrient fuel. During digestion, carbohydrate is broken down into glucose. Glucose circulates in the blood to be used for energy. If your muscles are deprived of glucose, your physical power suffers. You're low on gas and feel it. It becomes tough to stick to your eating program, and another attempt to lose weight could bite the dust.

But MCT oil can come to the rescue. You can still apply the low-carbohydrate dieting strategy, but without the corresponding loss of energy! Remember: MCT oil is burned in the body like a carbohydrate and spares glucose fuel for an energy-boosting effect.

Thus, by supplementing with MCT oil, you have a pure energy source to help prevent diet-induced fatigue. At the same time, MCT oil keeps your metabolism high, and a high metabolism is conducive to losing body fat.

How, then, should you plan your weight-loss diet to take advantage of MCT oil's powers? Here are some guidelines that explain how to substitute MCT oil for high-starch carbohydrate foods to facilitate fat-burning:

- Design a diet with the following percentages of nutrients: 20 to 25 percent protein, 40 to 50 percent carbohydrates, and the rest from fat, including MCT oil. Along with the restriction of carbohydrates, the higher percentage of protein in the diet helps control hormones, too (less insulin, more glucagon), in favor of fat loss.

- Eat higher-starch carbohydrate foods such as cereals, whole grains, potatoes, legumes, bread, pasta, or fruit only at breakfast, your midmorning snack, and lunch to adequately fuel your daily activities. (For carbohydrates, stick to natural ones such as potatoes and whole grains. Natural carbohydrates are used more efficiently by the body than their processed counterparts—breads and pastas, for example—because they are preferentially stored as glycogen, rather than as fat, if not used first for energy.)

- Do not eat any higher-starch carbohydrates after lunch.

- Use MCT oil (up to a tablespoon), taken with some nonstarchy vegetables for an afternoon snack. Have some additional MCT oil as salad dressing at dinner. Used this way, MCT oil moves your body into a fat-burning mode and helps speed up your metabolism.

- Stay on a diet plan like this no more than two weeks at a time.

Supplementing with MCT Oil

MCT oil should always be taken with food and can be drizzled over vegetables or made into salad dressings (see the recipes in chapter 12). You can also cook or bake with MCT oil just as you would with any other vegetable oil. Keep the heat at 350 degrees or lower, however, because MCT oil smokes at high temperatures. Don't store MCT oil in anything other than a glass container. It tends to soften containers made of certain types of plastic.

Gradually introduce MCT oil into your diet at the rate of a few teaspoons a day until you are eating two to three tablespoons a day. MCT oil contains 120 calories per tablespoon. This supplement is so rapidly absorbed that it tends to cause stomach cramping if too much is taken at one time or on an empty stomach.

Make sure you purchase pure MCT oil, not a product that's diluted. How can you tell? A good rule of thumb is that any MCT oil product that comes flavored is not the pure stuff. Read the label to see whether the product is cut with flavorings or other fillers.

Safety Considerations

MCT oil is generally safe for most people, except those with preexisting medical conditions. The supplement is not advised if you have a fatty liver or cirrhosis, because the fat is channeled directly into an already poorly working liver. Nor should you use MCT oil if you suffer from a chronic pulmonary disease such as emphysema or asthma, because it results in greater production of carbon dioxide—a side effect that can further complicate breathing. If you are diabetic, avoid MCT oil, too, because it promotes the production of ketones, by-products of fat-burning that are dangerous to people with this disease.

Worth mentioning, too: Extensive studies with animals have shown that MCTs do not have the potential to be carcinogenic.

MCT oil does not supply essential fatty acids, so make sure you're eating at least 2 teaspoons of such fats daily. Never substitute MCT oil entirely for essential fats in your diet, or else you'll risk an essential fatty acid deficiency.

Part V

Designing a Fat-Healthy Diet

· ELEVEN ·

A Smart-Fat Strategy

Without question, few nutrition topics have stirred up so much confusion as that of dietary fat. With conflicting recommendations on which fats to eat and which fats to avoid, it's easy to get overwhelmed. But don't despair. The first step is to give some thought to your personal health situation and decide what's best for you. For example: Do you have high cholesterol? Do you need to lose weight? Are you suffering from a health problem that might be treated with a fat supplement?

Once you've answered these questions, begin revamping your nutrition program to control both the amount and the type of fat you eat. There are some fats you'll want to include routinely in your diet, others you'll want to avoid, and still others you may want to take therapeutically, like medicine, on a temporary basis.

In this chapter there are several health scenarios. See where you fit in and note the recommendations. Once you've identified your niche, follow the recommendations, but tweak them where necessary. With nutrition, the key is to adopt healthy eating habits but to customize them to your lifestyle.

General Good Fat Guidelines for Everyone

By commandeering dietary fat—the type and the amount—you can stay healthy and delay or even prevent many life-threatening diseases. To help, the American Heart Association has developed dietary guidelines designed to promote good health and prevent disease. These guidelines are an excellent reference for deciding how much fat you need in your diet. Although mentioned elsewhere in this book, they bear repeating:

- Saturated fat should be held to 7 to 10 percent of your total calories (the number of calories you need to achieve and maintain a healthy body weight).

- Polyunsaturated fat intake should be up to 10 percent of your total calories.

- Monounsaturated fat should compose up to 15 percent of your total calories.

- Total fat intake should be no more than 30 percent of your total calories (600 calories a day based on a 2,000-calories-a-day diet). This guideline applies to total calories eaten over several days, such as a week.

If you're a math lover, you can calculate your own daily fat intake by using the following formulas:

Total fat:
Total calories per day × .30 = daily fat calories
Daily fat calories / 9 = _____ daily fat grams
(For example: 1,200 calories × .30 = 360 daily fat calories;
360 daily fat calories / 9 = 40 daily fat grams)

Saturated fatty acids (SFA):
Total calories per day × .07 = daily SFA calories
Daily SFA calories / 9 = daily SFA grams
(For example: 1,200 calories × .07 = 84 daily SFA calories;
84 daily SFA calories / 9 = 9 daily SFA grams)

Polyunsaturated fatty acids (PUFA):
Total calories × .10 = daily PUFA calories

Table 11

GRAMS OF FAT AND CALORIC LEVELS

Calorie Level	Saturated Fats (grams)	Polyunsaturated Fats (grams)	Monounsaturated Fats (grams)	Total Fats (grams)
1,200	9	13	20	40 or less
1,500	12	16	25	50 or less
1,800	14	20	30	60 or less
2,000	16	22	33	67 or less
2,200	17	24	36	73 or less
2,500	19	28	42	83 or less
3,000	23	33	50	100 or less

Daily PUFA calories / 9 = daily PUFA grams
(For example: 1,200 calories × .10 = 100 daily PUFA calories;
100 daily PUFA calories / 9 = 13 daily PUFA grams)

Monounsaturated fatty acids (MUFA):
Total calories × .15 = daily MUFA calories
Daily MUFA calories / 9 = daily MUFA grams
(For example: 1,200 calories × .15 = 100 daily MUFA calories;
100 daily MUFA calories / 9 = 20 daily MUFA grams)

Table 11 shows how these recommendations and calculations translate into actual grams of fat.

Once you've calculated your own dietary allowance for total fat and saturated fat, be sure to read food labels for the fat content per serving. The grams of fat are listed under *nutrition facts* on any food package that provides a nutrition label.

Another way to monitor your fat intake is by limiting the majority of foods in your diet to those that have only 30 percent or less of their calories from fat. By using the information on the nutrition label, you can easily determine if a food meets this criterion. Use the following formula to find percentage of calories from fat:

$$\text{Percent calories from fat} = \text{Calories of fat per serving} / \text{total calories per serving} \times 100$$

Suppose an item of food has 36 fat calories in a serving, and the total calories per serving are 220. Here's an example using that label information:

$$36 \text{ fat calories}/220 \text{ calories per serving} \times 100 = 16\% \text{ of calories from fat}$$

In addition to controlling the amount of fat you eat, you must make good fat choices. Some of this information has been covered in previous chapters, but here's a quick summary for review:

INCREASE YOUR INTAKE OF OMEGA-3 FATTY ACIDS. Eat cold-water fish two to three times a week. Fish servings should be at least 3 ounces—about the size of a deck of cards. In addition, use flaxseed in recipes and meals, or eat a tablespoon of flaxseed oil each day. Flaxseed is the richest plant source of alpha-linolenic acid, an omega-3 fatty acid.

PUT MORE MONOUNSATURATED ACIDS IN YOUR DIET. Use extra-virgin olive oil in salad dressings, for bread dipping, or for cooking. Canola oil is another fat high in monounsaturated fats that makes an excellent oil for cooking or baking. In addition, use avocados in salads; they are chock-full of healthy monounsaturated fats. Also loaded with these good fats are nuts. Sprinkle them on cereal, yogurt, salads, and other foods.

CUT BACK ON OMEGA-6 FATS. These include vegetable oils such as corn, cottonseed, safflower seed, sunflower seed, and soybean oils—and products like margarines that are made from them. For margarines, opt for spreads made with canola oil instead. Processed foods are also high in omega-6 fats and should thus be limited in your diet.

CURTAIL YOUR SATURATED FATS. These are found mostly in fatty cuts of meat, poultry skin, whole and 2-percent milk, cheese, butter, many ice creams, and coconut oil. Here are some easy ways to control your intake of saturated fats:

• Choose lean cuts of good- or choice-grade meat like round, sirloin, and flank, and eat portions that are no larger than the palm of your hand.

Chicken, turkey, and fish are always leaner meat choices, or substitute fish a couple of times during the week.

- When preparing and eating meats, make sure to trim all visible fat and skin, and use cooking racks to bake, broil, grill, steam, or microwave, to avoid melting the fat back into the meat.

- When eating lunch meats, select low-fat or fat-free chicken or turkey breast, rather than high-fat bologna or salami.

- Choose low-fat or skimmed products rather than whole milk, and include them two to three times each day.

- Eat reduced-fat or fat-free cheese.

ELIMINATE OR LIMIT TRANS-FATS. Found in processed foods and fast foods, these fats are the "baddest" of the bad, linked to heart disease and many other life-threatening illnesses. Check the labels of foods and select products without partially hydrogenated oils in the list of ingredients. Switch to margarines that are trans-free. Also, nix the fried foods at restaurants. Chances are, they're prepared in oils full of trans-fats.

Guidelines for High Cholesterol

To bring your cholesterol down, you need to reduce your intake of saturated fat and dietary cholesterol. The American Heart Association recommends that you:

- Restrict your fat intake to 30 percent of your total calories. Less than 7 percent of your total calories should come from saturated fats and trans-fats.

- Limit your daily intake of dietary cholesterol to less than 200 milligrams.

- Select nonfat dairy products.

- Eat no more than 6 ounces a day of lean meat, fish, or skinless poultry.

- Restrict your use of fats and oils to five to eight servings a day. (For examples of servings, see Table 12.)

- Eat five or more daily servings of fruits and vegetables, and six or more servings of grains—bread, cereal, or rice, for example.

Beyond these recommendations, there are other measures worth taking. For example, two or three times a day, try substituting butter and margarine with cholesterol-lowering spreads such as Benecol or Take Control. Both have been clinically proven to reduce total cholesterol and LDL cholesterol.

In addition, eat flaxseed as a regular part of your diet, supplement with evening primrose oil (3 to 4 grams a day), and use olive oil in your cooking as part of your allowable daily fat portion.

Also, be careful to not overindulge on sugary foods. Research has shown that sugar may be a risk factor for heart disease, possibly because it generates very-low-density lipoprotein (VLDL) cholesterol and triglycerides. Both are harmful to the heart. Thus, avoid products listing more than 5 grams of sugar per serving on the label. If the specific amount of sugar is unlisted, shun products with sugar listed as one of the first four ingredients on the label. Sugar goes by various other names too: sucrose, dextrose, maltose, lactose, maltodextrin, corn syrup, to name just a few.

In addition to dietary measures, begin a regular exercise program—and stick with it. Exercising can lower your cholesterol by about 9 percent and raise your HDL cholesterol by 5 to 15 percent. The key is to do at least thirty minutes of exercise most days of the week.

If your cholesterol doesn't drop with diet and exercise, consult your physician. You may need a cholesterol-lowering medication.

Guidelines for High Triglycerides

Triglycerides are the chemical form in which fat exists in food, as well as in body fat. They're also present in blood plasma, and together with cholesterol, form the lipids in your blood.

High blood triglycerides are most often associated with heart disease and may be a consequence of type II diabetes. Table 13 lists the various measurements for triglycerides.

Several factors can elevate your triglyceride levels: drinking alcohol, taking estrogen, and poorly controlling your diabetes. Higher-carbohydrate diets

Table 12

CONTROLLING CHOLESTEROL:
SERVINGS AND SELECTIONS FOR FATS AND OILS

Servings per day	Serving Size
5 to 8 servings	1 tsp. vegetable oil or regular margarine 2 tsp. diet margarine 1 tbsp. salad dressing 2 tsp. mayonnaise or peanut butter 3 tsp. seeds or nuts ⅛ medium avocado 10 small or 5 large olives
Choose from:	Vegetable oils or margarines with no more than 2 grams of saturated fat per tablespoon: canola, corn, olive, safflower, sesame, soybean, sunflower, and walnut; and almond, avocado, and hazelnut monounsaturated oils. For margarines, choose those that contain liquid vegetable oil as the first ingredient. For mayonnaise and salad dressing, select those with no more than 1 gram of saturated fat per tablespoon.

Source: American Heart Association.

tend to increase triglycerides, too. So does eating too much high-fructose corn syrup, a refined version of fructose made from corn and found in many processed foods and beverages.

Reducing your triglycerides requires a lifestyle fix in which your goals are to lose weight (if you're overweight), reduce the cholesterol and saturated fat in your diet, avoid or limit alcohol consumption, and become more active.

Also urged by medical experts is substituting monounsaturated fats and polyunsaturated fats for saturated fats. With this approach, 40 percent of your total daily calories should come from fat. However, half of that fat should come from monounsaturated fats, which have been shown in research to lower triglyceride levels. A higher monounsaturated fat diet might include olive oil or canola oil as part of your daily diet. When you increase

Table 13

TRIGLYCERIDE READINGS

Normal triglycerides	Less than 200 mg/dl
Borderline-high triglycerides	200 to 400 mg/dl
High triglycerides	400 to 1,000 mg/dl
Very high triglycerides	Greater than 1,000 mg/dl

Source: American Heart Association.

your calories from fat, be careful not to increase your total calories, particularly if you're trying to watch your weight. In addition, reduce your intake of carbohydrates because they elevate triglycerides and decrease HDL cholesterol.

Fat supplements that may help regulate triglycerides include perilla oil and DHA.

Guidelines for High Blood Pressure

A low-fat diet designed to control cholesterol is also beneficial for reducing high blood pressure (hypertension), particularly if it includes plentiful amounts of fruits, vegetables, whole grains, and low-fat dairy products. This type of diet is rich in nutrients such as calcium, potassium, and magnesium, all of which help normalize blood pressure.

One mineral that should be restricted, however, is sodium, found in table salt. Too much salt in the body tends to narrow the diameter of blood vessels. When this happens, the heart has to work harder to pump the same amount of blood, and blood pressure soars as a result. Excessive salt also makes the body retain too much water, and this may cause a rise in blood pressure.

If you suffer from high blood pressure, do not add salt to your food and avoid salty foods such as snack food, smoked meats, pickled foods, cheese and cheese products, fast foods, and canned foods. If you miss the taste of salt on foods, try a salt substitute or experiment with various herbs and spices when cooking.

Healthy fats that may help drive blood pressure down include fish oil supplements (and fish), borage oil, black currant oil, and olive oil.

Losing weight helps normalize blood pressure too. On average, you can cut your blood pressure by several points by losing weight through diet and exercise.

Another way to reduce high blood pressure, is with aerobic exercise, such as walking, jogging, running, swimming, bicycling, and so forth. Most studies of hypertensive people show that a reduction can occur with as little as three exercise sessions a week for thirty to sixty minutes each time.

If your blood pressure can't be controlled by lifestyle changes, you'll probably need to take medication prescribed by your doctor.

Guidelines for Weight Loss

Being overweight, defined as 20 percent or more above your ideal weight, puts you in harm's way of numerous life-threatening diseases. Among them: heart problems (overweight increases LDL cholesterol and lowers HDL cholesterol), stroke, high blood pressure, and diabetes.

Without question, it can be challenging to lose weight, especially if you're way off the ideal. But it's not impossible either.

The first step is to eat fewer calories than your body uses each day. To lose a pound of body fat, you have to create a 3,500-calorie deficit, either by eating less, exercising more, or both. By cutting your total calorie intake by 500 calories each day, for example, you should be able to lose one pound a week (500 calories×7 days)—a safe rate of weight loss. If you add exercise to this equation and burn any extra calories, your fat loss will be even greater. An hour of exercise, for example, can burn up anywhere from 250 to 500 calories.

Your weight-loss diet should be as low in fat as possible, since reducing dietary fat is one of the best ways to shed pounds. By keeping your total fat intake to 20 percent of total daily calories, you may be able to lose body fat with less restriction in total calories.

In addition, try to curb your intake of fat-forming foods such as sugar, processed foods, and alcohol. By limiting these foods, you'll automatically reduce the number of calories in your diet.

Two fat supplements that may help you lose body fat are conjugated linoleic acid (3 grams daily) and medium-chain triglyceride oil (MCT oil).

Try substituting MCT oil for some of your fat calories each day and for a portion of your higher-starch carbohydrates. MCT oil revs up your metabolism to help your body burn energy more efficiently.

In addition to the diet tips, many people have successfully lost weight by following a higher-fat, moderate-protein, and restricted-carbohydrate diet. Generally, these diets work by encouraging your body to switch from burning stored carbohydrate (glycogen) for energy to burning stored body fat. One of the most popular low-carbohydrate diets is the Atkins Diet.

In a study of the diet conducted at the Durham VA Medical Center in North Carolina, mildly obese people lost about twenty-one pounds in four months on the diet. What's more, they showed a 6.1-percent drop in cholesterol, a 40-percent decline in triglycerides, and an increase in HDL cholesterol by about 7 percent.

The diet restricted carbohydrate intake to less than 20 grams daily and included vitamin supplements, fish oil supplements, and twenty minutes of exercise at least three times a week.

Not all diets work equally well for everyone. You should choose a weight-loss diet that fits your food preferences and lifestyle. Further, it should be a diet you're willing to stick with for the long haul, in order to lose the required amount of weight.

Guidelines for Diabetes

Diabetes is a complex disease, requiring treatment that involves diet, exercise, lifestyle changes, and, for many people, injectable insulin or oral diabetes drugs. Generally, the recommended diet is one similar to that for high cholesterol: a low-saturated fat, low-cholesterol, high-fiber diet. In addition, a number of good fats may be therapeutically beneficial for treating diabetes: flaxseeds and flaxseed oil, black currant oil, evening primrose oil, and, possibly, DHA. Fish oil supplements are generally not recommended if you have diabetes. Rely on fish to get your dose of its essential fatty acids.

High triglycerides tend to occur simultaneously with high LDLs and low HDLs. If your diabetes is complicated by high triglycerides (greater than 250 mg/dl), the American Diabetes Association suggests that you try a higher-fat diet in which 40 percent of your total daily calories is derived from fat. However, half of your fat should come from monounsaturated fats, which have been shown in research to lower triglyceride levels.

Guidelines for Arthritis

Arthritis, a serious and potentially crippling disease, affects millions of people worldwide. It attacks the joints as well as the muscle and connective tissues surrounding them. There are more than 100 different forms of arthritis. The two most common are osteoarthritis, caused by wear and tear on the joints, and rheumatoid arthritis, an autoimmune disease in which the body's immune system attacks itself.

To some extent, supplements can play a role in treating arthritis. Omega-3 fatty acids such as fish oil and flaxseed oil are helpful in relieving symptoms. So are GLA-rich supplements such as borage oil, black currant oil, and evening primrose oil. Another good move is to reduce your intake of saturated fat such as red meat, dairy products (whole milk, 2-percent milk, ice cream, butter, cheese, cream), margarine, shortening, lard, cocoa butter, and fried foods. The less saturated fat you eat, the better omega-3 fats can work their healing magic.

Therapeutic Use of Essential Fatty Acid Supplements

The essential fatty acid supplements discussed in this book are simply nutrients, extracted from food or plants, that have a much gentler effect on the body than do prescription drugs. That being so, you may want to take certain supplements from time to time, on a therapeutic basis, to help alleviate symptoms or coax your body to heal naturally. In certain circumstances— say a migraine or menstrual pain—it's usually wise to try the gentler agent first, before resorting to prescription or over-the-counter drugs. Do so with the okay of your physician or psychiatrist, however, particularly since essential fatty acid supplements contain factors that may thin your blood, which can increase bleeding times.

You may experience some other mild side effects from taking essential fatty acid supplements, including upset stomach, burping, or minor bowel problems such as loose stools. These can be minimized by taking the supplement with meals and in divided doses throughout the day.

As fats, essential fatty acid supplements contain calories. Typically, there are about 10 calories in a 1,000-milligram softgel, so you'll need to take this into consideration if you're counting calories.

Table 14

TREATING DISEASE WITH ESSENTIAL FATTY ACID SUPPLEMENTS

Disorder/Illness	Supplements	Dosage	Comments
Abnormal clotting	Flaxseed oil Perilla oil	40 grams daily 6 grams daily	Check with your physician because EFA supplements may thin your blood, or interfere with the action of blood-thinning medication.
ARDS	Borage oil	Consult your physician regarding dosage.	Supplementation should support conventional medical treatment under the care of a physician.
Arrhythmias	Fish oil	4 grams daily	Supplementation should support conventional medical treatment under the care of a physician.
Asthma	Perilla oil	6 grams daily	Supplementation should support conventional medical treatment under the care of a physician.
Brain diseases	DHA	100 milligrams daily if you eat fish; 200 milligrams daily if you eat no fish.	Supplementation should support conventional medical treatment under the care of a physician.
Breast pain	Evening primrose oil	6 grams daily	Supplementation often works better than standard drug treatment.

Table 14 (continued)

TREATING DISEASE WITH ESSENTIAL FATTY ACID SUPPLEMENTS

Disorder/Illness	Supplements	Dosage	Comments
Depression	Fish oil	6 grams daily	Supplementation should support conventional medical treatment under the care of a physician or psychiatrist.
Diabetes	DHA may be helpful, although research is preliminary.	Correct dosage for treating diabetes is unknown.	Take supplements only after consultation with your physician and dietitian.
Diabetic nerve damage	Evening primrose oil	1–4 grams daily	Supplementation should support conventional medical treatment under the care of a physician.
Eczema	Evening primrose oil	1–4 grams daily	One of the best natural treatments available for eczema.
High cholesterol	Evening primrose oil	1–4 grams daily	Should be used in conjunction with a cholesterol-lowering diet that includes reduced intake of saturated fats.
High triglycerides	Fish oil	100 milligrams daily if you eat fish; 200 milligrams daily if you eat no fish.	Should be used in conjunction with a low-fat diet in which 40 percent of fat calories are derived from monounsaturated fats.

Table 14 (continued)

TREATING DISEASE WITH ESSENTIAL FATTY ACID SUPPLEMENTS

Disorder/Illness	Supplements	Dosage	Comments
High triglycerides	DHA CLA Perilla oil	3 grams daily Up to 3 grams daily 6 grams daily	Should be used in conjunction with a low-fat diet in which 40 percent of fat calories are derived from monounsaturated fats.
Hypertension	Fish oil Borage oil	3 grams daily 3 grams daily	Supplementation should support conventional medical treatment under the care of a physician. Borage oil should not be taken with anticonvulsants.
Inflammatory bowel disease	Fish oil	3 grams daily	Supplementation should support conventional medical treatment under the care of a physician.
Menstrual disorders	Evening primrose oil	1–4 grams daily	One of the most effective natural treatments for a variety of menstrual problems, including cramps, premenstrual syndrome (PMS), breast pain, and heavy bleeding.
Migraines	Evening primrose oil	1–4 grams daily	Relieves inflammation associated with migraine pain.
Obesity	CLA	3–4 grams daily	Should be used in conjunction with a low-fat, reduced-calorie diet, plus regular exercise.

Table 14 (continued)

TREATING DISEASE WITH ESSENTIAL FATTY ACID SUPPLEMENTS

Disorder/Illness	Supplements	Dosage	Comments
Obesity	MCT oil	1–2 tablespoons daily to replace some of daily fat portion	Should be used in conjunction with a low-fat, reduced-calorie diet, plus regular exercise.
Osteoarthritis	Lyprinol (green-lipped–mussel supplement)	210 milligrams daily	Supplementation should support conventional medical treatment under the care of a physician.
Psoriasis	EPA	1.8 grams daily	Supplementation should support conventional medical treatment under the care of a physician.
Rheumatoid arthritis	Black currant oil	525 milligrams daily	Supplementation should support conventional medical treatment under the care of a physician.
	Borage oil	1.4 grams daily	
	Fish oil	3–5 grams daily	

Many manufacturers make essential fatty acid supplements that contain a blend of various omega-3 and omega-6 fats. These may be worth a try to ensure that you get a good balance of the right fats.

Table 14 reviews the various disorders that can be treated with essential fatty acid supplements and provides the dosages generally recommended by health care practitioners.

• T W E L V E •

Good Fat Cooking

There's certainly no better way to start reaping the benefits of good fats than to start cooking with them. Using healthful fats and oils such as MCT oil, flaxseeds and flaxseed oil, olive oil, canola oil, and sesame seeds and sesame oil in recipes is an effortless way to get a healthy dose of essential fats, monounsaturated fats, and other good-for-you fats. With that in mind, here are some recipes that incorporate many of the fats discussed in this book.

MCT Oil Recipes

Medium-chain triglyceride (MCT oil) is a special type of dietary fat that is metabolized quickly and not easily stored as body fat. It is a useful supplement if you are trying to trim down and get in shape. You can also cook or bake with MCT oil just as you would with any other vegetable oil. Keep the heat at 350 degrees or lower, however, because MCT oil smokes at high temperatures.

The following MCT oil recipes are used with permission from Parrillo Performance, 4690K Interstate Drive, Cincinnati, Ohio 45246, 800-344-3404. This company makes an unflavored MCT oil product called CapTri®.

Fried Chicken

Makes 5 servings
438 calories per serving; 19 grams of fat per serving

> 2 cups shredded wheat (crumbled very fine)
> ¼ cup oat bran
> 1 tsp. onion powder
> ½ tsp. garlic powder
> ½ tsp. barbecue seasoning
> ½ tsp. coarse black pepper
> 1 tsp. Mrs. Dash Lemon and Herb seasoning
> 2 pounds chicken breasts
> 5 tbsp. MCT oil (unflavored)

1. Place chicken in medium-sized glass bowl and coat thoroughly with MCT oil. Set aside.

2. In another bowl, mix all other ingredients for breading. Dip chicken one piece at a time into the breading mixture and toss until well coated.

3. Heat skillet (moderate heat) with any remaining MCT oil from the chicken dip. Reduce heat to low and cover. Turn chicken breasts occasionally to cook evenly on all sides. Chicken is done when breading is brown and meat is white, juicy, and tender. Be careful to not overcook, as this will dry out your chicken and make it tough.

Cod Fillet Italiano

Makes 4 servings
350 calories per serving; 10 grams of fat per serving

> 2 lbs. cod fillets
> 1 cup oatmeal flour
> 3½ cups chopped tomato, blended in blender
> 3 tbsp. minced parsley
> 1 tsp. oregano
> ¼ tsp. garlic powder
> ¼ tsp. onion powder
> ⅛ tsp. pepper
> 2 tbsp. MCT oil (unflavored)

Preheat oven to 325 degrees. In a small bowl, combine oatmeal flour, parsley, and other seasonings with 1 tbsp. MCT oil. Spread remaining MCT oil in the bottom of a 9-inch baking dish. Spread oatmeal flour mixture evenly over the fillets. Bake for 20 minutes. Remove from oven and pour blended tomatoes over fillets. Bake for an additional 10 to 15 minutes, or until fish flakes when tested with a fork.

Mexican Black Bean and Turkey Salad

Makes 5 servings
485 calories per serving; 12.5 grams of fat per serving

¾ tsp. cumin

¾ tsp. chili powder

⅛ tsp. salt

⅛ tsp. ground pepper

1 lb. turkey breast, cut into strips

1 cup curly endive

1 cup romaine lettuce

½ cup fresh orange segments

¼ cup chopped red onion

2½ cups black beans, washed and drained

½ cup cilantro

¼ cup lime juice

3 tbsp. MCT oil (unflavored)

Extra MCT oil for cooking

⅛ tsp. salt

1 small garlic clove, chopped

1 tbsp. fresh orange juice

1. Place cumin, chili powder, salt, and ground pepper in a bag and shake to mix. Add turkey to bag and continue shaking to coat turkey. Let marinate for 1 to 3 hours. Sauté turkey in skillet lightly coated with MCT oil until golden brown on the outside.

2. Mix endive and romaine lettuce with orange segments, onions, black beans, and cilantro. Add turkey.

3. For dressing, mix lime juice, MCT oil, salt, garlic, and orange juice. Toss with vegetable and turkey mixture.

White Chili

Makes 6 servings
450 calories per serving; 16 grams of fat per serving

2½ cups canned white beans
1 quart water
1 quart chicken stock
1 cup chopped onions
3 chopped garlic cloves
½ chopped green chili peppers
2 tsp. cumin
1½ tsp. crushed oregano
1 tsp. coriander
¼ tsp. cloves
¼ tsp. cayenne pepper
1 lb. baked turkey breast, chopped
6 tbsp. MCT oil (unflavored)

1. Combine beans, water, stock, half of the onions, and garlic in a large pot and bring to a boil. Reduce heat, cover, and simmer until onions are soft.

2. Heat 3 tbsp. MCT oil in a skillet. Add remaining chopped onion, chili peppers, and spices. Cook until tender. Add this mixture, plus cooked turkey, to the pot and cook 30 minutes on moderate heat.

Home Fries

Makes 3 servings
266 calories per serving; 14 grams of fat per serving

3 medium potatoes, sliced
3 tbsp. MCT oil (unflavored)
½ tsp. onion powder
½ tsp. garlic powder
Dash of red pepper
¼ cup water

1. Place all ingredients except water in a large glass bowl and toss until potatoes are evenly coated with oil and spices. Place in a hot nonstick skillet, cover, and cook on medium heat for about 5 minutes.

2. Pour water in skillet and turn potatoes with spatula. Cover again and cook until potatoes are tender and lightly browned, stirring occasionally.

Fried Squash

Makes 6 servings
360 calories per serving; 10.8 grams of fat per serving

> 2 cups summer squash, cut into strips
> 1 egg white
> Chili powder to taste
> Pepper to taste
> 1½ cups oatmeal, ground fine in the blender
> 1/3 cup water
> 4 tbsp. MCT oil (unflavored)

Preheat oven to 425 degrees. Mix all ingredients together except squash strips to form a batter. Dip strips into batter to coat. Spray cookie sheet lightly with vegetable oil spray. Place strips on cookie sheet. Bake 10 to 15 minutes or until browned.

Stuffed Mushrooms

Makes 4 servings
187 calories per serving; 11 grams of fat per serving

> ¾ pound ground turkey
> 4 cups medium-sized fresh mushrooms
> ½ cup shredded wheat, crumbled
> 1 tsp. Mrs. Dash seasoning
> 2 tbsp. MCT oil (unflavored)
> Parsley to taste

1. Place MCT oil in a frying pan. Add turkey and brown over medium heat. Add shredded wheat and Mrs. Dash seasoning. Cook for 5 minutes, then remove pan from heat.

2. Remove stems from mushrooms. Wash mushroom caps thoroughly and place on a cookie sheet that has been sprayed with vegetable spray.

3. Spoon turkey mixture into mushroom caps and bake for 10 minutes, or until mushrooms are brown and tender. Garnish with parsley.

MCT Raspberry Vinaigrette

124 calories per tablespoon

> 1 cup MCT oil (unflavored)
> ¾ cup raspberry vinegar
> 1 tbsp. Dijon mustard
> 1 tbsp. honey
> 1 tbsp. shallots, minced

Mix ingredients completely in a glass container with a tight-fitting lid. Shake well and serve over salads.

Garlic Lover's MCT Oil Pesto

Makes 1 serving
134 calories per serving; 14 grams of fat per serving

> ⅓ cup chopped fresh basil
> ¼ cup minced garlic
> 2 tbsp. chopped onion
> 1 tbsp. MCT oil (unflavored)
> Pinch of salt if desired

Sauté all ingredients in a small nonstick frying pan. Pesto can be served over rice, pasta, or chicken.

Mexican Bean Dip

> 1¼ cup cooked pinto beans, washed and drained
> 1 large tomato, finely chopped
> 4 tbsp. MCT oil (unflavored)
> ½ tsp. chili powder
> ½ tsp. cumin

Purée beans in a food processor or blender, or mash with a fork. Add rest of ingredients and continue to blend.

Corn Chips

½ cup boiling water

2 tbsp. MCT oil (unflavored)

1 cup cornmeal

¼ tsp. chili powder

Popcorn salt

Preheat oven to 350 degrees. Pour water over MCT oil, cornmeal, and chili powder. Mix well with a fork. Shape dough into small one-inch balls and place far enough apart on a nonstick cookie sheet so that they do not touch when pressed flat. Press balls as flat and as thinly as you can, shaping them into triangles, ovals, or rectangles. Sprinkle lightly with pinches of popcorn salt. Bake about 30 minutes or until edges start to brown. Chips should be thin and crisp. Serve with Mexican Bean Dip.

Potato Chips

2 medium potatoes, with or without skin

1 tbsp. MCT oil (unflavored)

Pinch of garlic powder

Popcorn salt to taste

Preheat oven to 325 degrees. Slice potatoes very thinly—thin enough so that they are almost transparent. Soak slices in MCT oil and garlic powder for 15 minutes. Place potato slices on a nonstick 14×10-inch baking sheet. Do not overlap them. Sprinkle with pinches of popcorn salt. Bake for 30 minutes, or until edges and smaller slices are browned. Remove from cookie sheet and drain on a paper towel.

Variation: Try flavoring your chips with barbecue seasoning powder instead of garlic.

Olive Oil Recipes

From olives and olive oil come some of the most nutritious substances on the planet, namely heart-healthy monounsaturated fatty acids and antioxidants. Olive oil is a terrific cooking oil because it can withstand higher cooking temperatures and is slow to oxidize, with little production of disease-causing free radicals.

The following olive oil recipes, rich in not only monounsaturated fats but also in omega-3 fats, are used with permission from the California Olive Industry, 1903 North Fine #102, Fresno, California 93727, 559-456-9099, www.calolive.org.

Salmon with Pine Nut–Rosemary Olive Crust

Makes 4 servings

475 calories per serving; 31 grams of fat per serving (16 grams of monounsaturated fat)

> ¾ cup ripe olives
>
> ½ cup pine nuts (filberts or pecans may be substituted for pine nuts)
>
> 1 tbsp. fresh chopped rosemary
>
> 4 (6–8 oz.) portions of salmon or halibut fillets/steaks
>
> To taste, salt and pepper

1. In a food processor bowl fitted with a blade, pulse olives until finely chopped; transfer to a shallow bowl. Pulse pine nuts in a food processor bowl until minced; transfer to olive bowl. Blend rosemary into pine nut–olive mixture.

2. Season salmon with salt and pepper, if desired. Press ½-cup portions of olive mixture onto the surface of each fish fillet. Bake on spray-coated tray in 450 degree oven for 15 to 20 minutes, or until fish is firm.

Shellfish Fettuccine with Ripe Olives and Garlic

Makes 8 servings

430 calories per serving; 18 grams of fat per serving (11 grams of monounsaturated fat)

> ¼ cup olive oil
>
> 3 pounds Roma tomatoes, diced
>
> ½ cup capers
>
> ¼ cup anchovies, chopped
>
> 2 tbsp. garlic, chopped
>
> 1 (6-oz.) can California Ripe Olives, whole, pitted
>
> 3 pounds shrimp, medium, peeled, deveined

2 (1-lb.) packages fettuccine pasta, cooked, hot

2 tbsp. olive oil

½ cup flat-leaf parsley, fresh, chopped

Heat olive oil in shallow heavy pot. Add tomatoes, capers, anchovies, and garlic. Cook over medium heat until tomatoes release juices and mixture thickens, about 5 to 10 minutes. Add olives and shrimp. Simmer until shrimp is firm, about 4 to 5 minutes. Toss hot pasta with olive oil. Add shrimp mixture to hot pasta and toss well. Portion onto plates or shallow bowls. Sprinkle with chopped parsley.

California Black Olive Pesto

Makes 4 servings

362 calories per serving; 40.65 grams of fat per serving (27.65 grams of monounsaturated fat)

1 cup California Ripe Olives (pitted)

⅓ cup chopped fresh basil leaves

⅛ cup pine nuts

½ cup olive oil

¼ cup Parmesan cheese, grated

⅛ tsp. pepper

Place olives, basil, and nuts in a food processor and pulse, until finely chopped. Slowly add olive oil while still pulsing. Add Parmesan, plus pepper.

White Bean and Ripe Olive Gratin

Makes 8 servings

321 calories per serving; 17 grams of fat per serving (11 grams of monounsaturated fat)

2 tbsp. olive oil

1 cup celery, thinly sliced

½ cup red onion, thinly sliced

1 tsp. garlic, minced

2 cups each: Roma tomatoes, seeded and diced; and zucchini, ¼-inch sliced

2 cups California Ripe Olives, sliced

¼ cup fresh sage, chopped

2 (15-oz.) cans white beans, rinsed

1 cup bread crumbs, fresh

1 tsp. garlic

¼ cup parsley, chopped

1 tsp. lemon zest, grated

2 tbsp. olive oil

1. Preheat oven to 350 degrees. Heat olive oil in heavy pot. Add celery, onions, and garlic. Sauté over medium-high heat for 3 minutes. Add tomatoes and zucchini and simmer for 5 minutes. Remove from heat. Add olives and sage. Purée about ¼ of the beans and add all beans to the tomato mixture. Mix well and adjust seasoning with salt and pepper.

2. Transfer to a buttered 2-quart shallow baking dish. Combine the last five ingredients in a small bowl. Mix well and sprinkle evenly over casserole. Bake at 350 degrees until bubbly and golden, about 45 minutes. Let rest 5 or 10 minutes prior to serving.

Mediterranean Baked Zucchini

Makes 6 servings

130 calories per serving; 11 grams of fat per serving (7 grams of monounsaturated fat)

1 cup California Ripe Olive wedges

2 tbsp. chopped parsley

1 tbsp. chopped scallion greens

1 tbsp. lemon juice

6 Roma tomatoes (sliced)

3 medium zucchini (sliced)

2 tbsp. olive oil

½ tsp. salt

¼ tsp. black pepper

1. Combine first four ingredients in a small bowl. Layer tomato and zucchini slices in a casserole dish sprayed with nonstick spray. Sprinkle olive oil mixture over the top. Sprinkle with olive oil, salt, and pepper.

2. Bake in a preheated 350-degree oven until zucchini is tender, about 15 to 20 minutes.

Olive Caesar Salad

Makes about 14 cups salad; 8 servings
144 calories per serving; 13 grams of fat per serving (8 grams of monounsaturated fat)

Garlic croutons (recipe follows)
1 large egg
¼ cup olive oil
1 clove garlic, minced
12 cups rinsed and crisped bite-size pieces romaine lettuce
Pepper
2 tbsp. lemon juice
¾ cup California Ripe Olives, wedged
3 to 4 canned anchovy fillets, chopped
¼ cup grated Parmesan cheese

1. Prepare garlic croutons; set aside. Immerse egg in boiling water to cover for exactly 1 minute; remove from water and set aside. Beat to blend oil and garlic in a large serving bowl. Add lettuce with a few croutons and pepper; mix gently but thoroughly. Break coddled egg over salad and sprinkle with lemon juice; mix well. Add olives, anchovies, and cheese; mix again. Add remaining croutons; mix gently. Serve immediately.

2. Garlic croutons: In a 9-inch pan, combine 1 tablespoon olive oil with 1 clove garlic, minced. Add 1 cup ¾-inch cubes day-old French bread and mix to coat. Bake bread cubes in a 325-degree oven until crisp and tinged with brown, 15 to 20 minutes; stir occasionally.

Couscous Salad with Ripe Olives and Roasted Vegetables

Makes 8 servings
375 calories per serving; 16.5 grams of fat per serving (10.5 grams of monounsaturated fat)

2 cups California Ripe Olives, halved
8 cups couscous, prepared, chilled
2 tsp. thyme, fresh, chopped
1 tsp. rosemary, fresh, chopped
½ cup red wine vinaigrette dressing
¼ cup capers
2 large zucchini, ½-inch lengthwise, sliced

3 large leeks, white only, lengthwise split

2 red bell peppers, seeded, quartered

10 garlic cloves, peeled

¼ cup olive oil

Preheat oven to 400 degrees. Combine first six ingredients in large bowl. Cover and set aside. Arrange zucchini, leeks, bell pepper, and garlic on roasting pan. Brush with olive oil and sprinkle as desired with salt and pepper. Roast in preheated oven until tender, about 35 to 40 minutes. Cool vegetables and cut into ½-inch dices; chop garlic. Add vegetables to reserved couscous mixture. Toss gently but well. Chill completely. Remove from refrigerator 30 minutes before serving.

Additional Good Fat Recipes

The following recipes are some of my personal favorites for using healthy ingredients such as flaxseeds, flaxseed oil, sesame seeds, sesame oil, and canola oil.

Natural Muffins

Makes 1 dozen muffins, which can be served at breakfast, for snacks, or at other meals
120 calories per muffin; 4 grams of fat per muffin

1 cup oatmeal flour

1 cup cornmeal

1 tbsp. baking powder

½ tsp. salt

3 egg whites

2 tbsp. honey

3 tbsp. canola oil

1 cup skim milk

½ cup flaxseeds

Mix dry ingredients (flour, cornmeal, baking power, and salt) together. In a separate bowl, blend the remaining ingredients and pour into the dry mixture. Blend thoroughly. Pour batter into muffin tins that have been sprayed with vegetable spray or into cupcake papers. Bake at 400 degrees for 20 to 25 minutes or until brown.

Low-Fat Flaxseed Apricot Bread

Makes 12 slices
104 calories per slice; 1.5 grams of fat

> 1 ½ cups oatmeal flour
>
> 1 tbsp. baking powder
>
> 1 tsp. baking soda
>
> ½ tsp. salt
>
> 1 tsp. cinnamon
>
> 1½ cups wheat bran
>
> 1 package dried apricots, cut into bits
>
> ½ cup unsweetened applesauce
>
> 1 cup boiling water
>
> ½ cup reduced-calorie maple syrup (fructose sweetened)
>
> 2 egg whites
>
> ½ cup flaxseeds
>
> 1 tsp. vanilla

Stir together oatmeal flour, baking powder, baking soda, salt, and cinnamon. In a separate bowl, combine bran, apricots, applesauce, and water. Stir well. In another bowl, blend egg whites, syrup, and vanilla. Add contents of both bowls to the dry mixture and blend well. Pour batter into a loaf pan that has been coated with vegetable spray. Bake for 1 hour in a 375-degree oven. Remove from oven and refrigerate. Slice bread when cool.

Quick Gourmet Flaxseed Oil Dressing

70 calories per tablespoon; 7.5 grams of fat per tablespoon

> ¼ cup white balsamic vinegar
>
> 3 tbsp. water
>
> 1 package (0.75 oz.) Good Seasonings Garlic & Herb salad dressing mix
>
> ½ cup flaxseed oil

Place vinegar and water in a container with a tight-fitting lid. Add salad dressing mix and shake vigorously until well blended. Add oil and shake again until well blended. Can be refrigerated for up to four weeks.

Sesame Cheese Hors d'Oeuvres

1 (8-oz). package fat-free cream cheese
Soy sauce
Sesame seeds

With a toothpick, poke holes in the cream cheese. Slowly pour soy sauce over the cream cheese so that the soy sauce flows into the holes. Roll the cream cheese in sesame seeds so that it is completely coated with seeds. Refrigerate. Serve with wheat crackers.

Mediterranean Salad

Makes 1 serving
543 calories per serving; 23 grams of fat per serving

½ cup garbanzo beans
2 oz. feta cheese
½ onion, chopped
½ red pepper, chopped
Lettuce
Quick Gourmet Flaxseed Oil Dressing

Arrange beans, cheese, onion, and red pepper on a bed of lettuce. Drizzle with 1 tbsp. of salad dressing.

Chicken Broccoli Orientale

Makes 4 servings
276 calories; 11 grams of fat per serving

2 tbsp. cornstarch
6 tbsp. soy sauce
1 package skinless chicken breasts (about 4), cut in cubes
¼ cup white wine
4 tsp. brown sugar
2 tsp. vinegar
2 tbsp. sesame oil
2 tsp. crushed red pepper
1 tbsp. chopped garlic

2 medium onions, cut into chunks
1 lb. bag frozen broccoli cuts

Blend cornstarch and 4 tbsp. soy sauce in a bowl. Add chicken and stir to coat. Mix the rest of the soy sauce with wine, brown sugar, and vinegar. Set aside. Heat oil in wok at 300 degrees and add red pepper and garlic. Cook for 1 minute. Add chicken. Stir-fry until chicken is cooked. Add onion and broccoli. Cover wok and cook mixture until broccoli is tender—about 5 minutes, stirring occasionally. Remove cover and add wine mixture. Cook until sauce becomes thickened—3 to 4 minutes. Serve over hot rice.

A Cooking Oil Primer

Today, more than any other time in culinary history, there is a huge array of oils that can be used in cooking. To enhance and maximize flavor, it helps to know which oils are best for which types of recipes. Table 15 provides an at-a-glance review of various cooking oils, their characteristics and uses.

Low-Fat Cooking

Perhaps for medical reasons, you have been told to reduce the amount of fat in your diet. Here are some tips for cutting fat in recipes, without sacrificing flavor:

- Trim the fat from any meat before cooking. The skin from poultry can be removed after cooking.

- Brown meats by broiling, grilling, or cooking in nonstick pans with little or no oil.

- Cook meat, poultry, and fish so that fat drains off.

- Use nonstick vegetable sprays because they reduce the amount of oil or shortening required for cooking.

- Marinate meat, fish, and poultry in low-fat or nonfat salad dressings for added flavor.

- Substitute low-fat milk for whole milk in recipes.

Table 15

USING COOKING OILS

Oil	Characteristics	Uses
Almond oil	Tastes slightly almond.	Making pastry.
Avocado oil	Smooth, rich taste.	Salad dressing, bread dipping, stir-frying.
Canola oil	Lowest of all oils in saturated fat; mild-tasting.	Salads, baking, deep-fat frying, stir-frying.
Corn oil	All-purpose oil.	Salad dressings, mayonnaise, deep-frying.
Flaxseed oil	Pungent flavor.	Salad dressings.
Hazelnut oil	Nutty flavor.	Salad dressings.
Olive oil	Comes in several forms.	Salad dressings, mayonnaise, bread dipping, cooking (but not baking).
Peanut oil	Mild flavor.	Stir-frying, deep-frying.
Rice oil	Sweet taste.	Salad dressings, cooking.
Safflower oil	Mild taste.	Salad dressings, stir-frying, all-purpose cooking.
Sesame oil	Comes in light (made from untoasted seeds) and dark (made from toasted seeds); nutty flavor.	Light version is good for frying, grilling, and in marinades. Dark version is good for flavoring foods, not for cooking.
Soybean oil	Light and mild-tasting.	Salad dressings, sautéing, deep-frying.
Sunflower oil	Bland flavor.	Salad dressings, all-purpose cooking.

Table 15 (continued)

USING COOKING OILS

Oil	Characteristics	Uses
Walnut oil	Light and mild-tasting; tastes like walnuts; goes rancid quickly.	Salad dressings.
Wheat germ oil	Nutty flavor, very high in vitamin E.	Salad dressings.

- Shun nondairy creamers and toppings, which are high in saturated tropical oils.

- Sauté vegetables in broth, bouillon, or wine, rather than in butter, margarine, or oil.

- Substitute low-fat yogurt, buttermilk, or low-fat cottage cheese for sour cream in recipes.

- Substitute the fat or oil with an equal amount of applesauce or fruit purée in baked products. Both are wonderful replacements for the fat in baked products such as quick breads.

- For sauces and dressings, use low-calorie bases such as vinegar, mustard, tomato juice, and fat-free bouillon, instead of high-calorie ones like creams, fats, oils, and mayonnaise.

- You can reduce the amount of fat or oil in most recipes by about a third without affecting the recipe.

- When making soups, stews, sauces, and broths, you can remove 100 calories of fat per tablespoon by chilling the liquid after cooking and skimming off the congealed fat.

- Use low-fat cheeses in place of regular cheeses in recipes.

- Use egg whites or egg substitutes to replace whole eggs in recipes.

- 3 tablespoons of cocoa powder, plus 1 tablespoon of vegetable oil, can replace 1 ounce of baking chocolate in recipes.

• G L O S S A R Y •

Acrylamide. A carcinogen that forms as a result of chemical reactions that take place during baking or frying at high temperatures. It is found in high amounts in potato chips, corn chips, frozen french fries, and fast-food french fries.

Advanced Glycation End Products (AGEs). Detrimental substances formed in the body when elevated blood sugar links up (or glycates) with proteins in the body. AGEs weaken bodily tissues, including blood vessels, and may set the stage for heart disease.

Aerobics. Continuous-action exercise that can be performed within the body's ability to use and process oxygen. Examples include walking, jogging, running, cycling, swimming, and cross-country skiing.

Alkylglycerols. A group of lipids similar in structure to triglycerides that are found in other fatty fish, as well as in human bone marrow and breast milk. Alkylglycerols have been scientifically studied since the 1930s for their ability to reduce radiation damage, suppress tumor growth, build blood, and accelerate wound healing.

Alpha-linolenic fatty acid (ALA). An omega-3 fatty acid found mostly in vegetables, nuts, seeds, and oils produced from those sources.

Antioxidant. A special class of nutrients that fight "free radicals," a group of cells that damage otherwise healthy cells.

Appendicitis. An inflammation and infection of the appendix.

Arachidonic acid (AA). A fatty acid found mostly in animal foods, but also synthesized from the omega-6 fat, linoleic acid, in a process involving enzymes. Arachidonic acid can be converted to bad prostaglandins and leukotrienes in the body.

Atherosclerosis. Narrowing and thinkening of the arteries caused by inflammation, or deposits of cholesterol, fats, and other substances.

Black currant seed oil. A rich source of alpha-linolenic acid and gamma-linolenic acid (GLA).

Borage oil. A highly concentrated source of gamma-linolenic acid (GLA).

Calories. Units that represent the amount of energy provided by food.

Cancer. A group of diseases characterized by the presence of cells that grow out of control.

Canola oil. A healthy monounsaturated fat extracted from the seeds of the rapeseed plant.

Carbohydrate Counting. A meal-planning system recommended for people with diabetes. This system estimates the number of carbohydrates in food and matches that amount to units of insulin.

Carbohydrates. A food group that serves as a major energy source for the body. Derived mostly from sugar and starch, carbohydrates are broken down into glucose during digestion and are the main nutrient that elevates blood glucose levels.

Cholesterol. A fatty substance found in some foods and manufactured by the body for many vital functions. Excess cholesterol and saturated fat can increase blood levels of cholesterol and can collect inside artery walls. This process contributes to heart disease.

Complex Carbohydrates. Carbohydrates (starches) made of multiple numbers of sugar molecules.

Conjugated linoleic acid (CLA). A naturally occurring fatty acid present in dairy products (most notably milk fat) as well as in meat, sunflower oil, and safflower oil. It is formed when the bacteria in a cow's gut breaks down the essential fatty acid, linoleic acid, in the food the animal eats. Marketed as a natural fat-loss supplement.

Constipation. A condition in which bowel movements occur less often than usual or consist of hard stools that are difficult to pass.

C-Reactive Protein (CRP). A protein that increases with the amount of inflammation in your coronary arteries. High levels of CRP are now believed to be the strongest and most significant predictors of heart disease, heart attack, and stroke. CRP can be measured by medical testing.

Cruciferous Vegetables. Vegetables that contain indoles, compounds that seem to protect against cancer. Broccoli, cauliflower, cabbage, and watercress are cruciferous vegetables.

Dementia. A condition in which a person gradually loses the ability to remember, think, reason, interact socially, and care for him- or herself. It is not a disease, but rather a cluster of symptoms triggered by diseases or conditions that adversely affect the brain. Alzheimer's disease is one example of a dementia.

Diabetes. A disease in which the body cannot produce enough insulin or cannot use insulin in a normal way. This leads to high levels of glucose in the blood.

Digestion. The breakdown of foods by enzymes so that nutrients can be absorbed by the body.

Dihomo-Gamma Linolenic Acid (DGLA). A fatty acid that is synthesized from gamma-linolenic acid (GLA) in an enzyme-controlled process. DGLA is converted to good prostaglandins in the body, but can also be converted to arachidonic acid.

Diverticulitis. The development of inflammation and infection in one or more diverticuli, which are bulges in the inner lining of the colon.

Diverticulosis. The formation of bulges (diverticuli) in the inner lining of the colon.

Docosahexaenoic acid (DHA). An important fatty acid constituent of the brain and retina. Fish is a rich source of DHA.

Duodenal Ulcer. A hole or break in the first part of the small intestine known as the duodenum.

Eicosanoids. Hormonelike substances that are produced in the body from fats. They include prostaglandins and leukotrienes.

Eicosapentaenoic acid (EPA). A very potent fatty acid that prevents platelets in the blood from abnormal clotting, and also helps reduce inflammation. Alpha-linolenic acid is converted to EPA in the body.

Energy Gels. A type of highly concentrated carbohydrates with a puddinglike consistency that are packaged in single-serve pouches. Energy gels are designed for athletes and exercisers.

Essential fatty acids (EFAs). Vitaminlike substances that have a protective effect on the body. They are called essential because the body cannot manufacture them. They must be obtained from food.

Evening primrose oil. A supplement made from the seeds of the evening primrose plant that is valued for its GLA content.

Fat replacers. Food additives concocted from carbohydrates, protein, and other fats to replace the fat in foods.

Fats. A food group that provides energy but is the most concentrated source of calories in the diet.

Fatty acid chain. Carbon atoms with hydrogen atoms attached and with an acid group at one end.

Fatty Acids. Components of either dietary fat or body fat.

Fiber. The nondigestible portion of plants that can lower fat and glucose absorption, assist in weight control, and promote a healthy digestive system.

Flax. A plant whose seeds yield a high amount of alpha-linolenic acid.

Free Radicals. Cellular aberrations, formed when molecules somehow come up with an odd number of electrons. These cells destroy healthy cells by robbing them of oxygen, and this robbery weakens the immune system.

Fructose. A simple sugar found in fruit and fruit juices.

Fructose Intolerance. A sensitivity to the fructose in fruit juices, sports drinks, or products containing high-fructose corn syrup, and sometimes to the natural fructose in fruit.

Galactose. A simple sugar that is a part of lactose.

Gallstones. Solid crystal deposits, usually made up mostly of cholesterol, that form in the gallbladder, an organ that is involved in digestion.

Gamma-linolenic acid (GLA). A fatty acid made from linoleic acid in the body. GLA has a number of benefits, including the ability to fight inflammation.

Ghrelin. A hormone produced mostly by cells in your stomach that triggers your desire to eat.

Glucagon. A hormone produced by the pancreas that opposes the action of insulin and helps liberate fat from storage.

Glucose. Blood sugar. It serves as a fuel for the body.

Glycemic Index of Foods. A system of rating foods according to how fast they elevate blood glucose. Foods lower on the scale—low-glycemic index foods—are sometimes recommended in nutritional therapy for diabetes and weight control.

Glycogen. Stored carbohydrates in the muscles and liver.

H. Pylori. Bacterium that live in the mucous membranes lining the digestive tract that are the most common cause of duodenal ulcers.

HDL (High-Density Lipoprotein). A type of cholesterol in the blood that has a protective effect against the buildup of plaque in the arteries.

Hemorrhoids. Swollen blood vessels in the anus, often the result of straining during a bowel movement.

Hempseed oil. A nutritional oil made from the seeds of the hemp plant. It is rich in omega-6 fats.

High-Fiber Diet. A food plan that supplies between 25 and 35 grams of fiber daily.

High-Fructose Corn Syrup. Made from corn starch, high-fructose corn syrup is a liquid comprised of roughly half fructose and half glucose and added to many processed foods, including soft drinks.

Hydrogenated fats. Polyunsaturated omega-6 fatty acids that have been synthetically altered in a process called hydrogenation in which hydrogen is forced into the oil.

Hydroxytyrosol. A powerful antioxidant in olives and olive oil that is technically classified as a polyphenol.

Hyperglycemia. Abnormally high levels of glucose in the blood.

Hypoglycemia. Low blood sugar.

Inflammation. A bodily immune response that is triggered when the body is under attack from germs and other invaders.

Insoluble Fiber. A type of fiber that supplies bulk to keep foods moving through the digestive system.

Insulin. A hormone that decreases blood glucose levels by moving glucose into cells to be used for fuel. It is also involved in protein synthesis and fat formation.

Insulin Resistance. A condition in which cells do not respond to insulin properly.

Insulinlike Growth Factors. Chemicals in the body that arise with increased insulin production. When produced in excess, they might promote cancer by increasing abnormal cell growth.

Irritable Bowel Syndrome. A condition characterized by alternating periods of diarrhea and constipation, often accompanied by cramping.

Lactose. The simple sugar found in milk.

Lactose Intolerance. A sensitivity to the simple sugar lactose in milk. It is caused by the lack of sufficient lactase, the enzyme required to digest lactose.

Lecithin. A phospholipid involved in the proper metabolism of cholesterol.

Leptin. A hormone produced in the body that acts like an appetite suppressant.

Leukotrienes. Substances in the body that are involved in inflammation. They are produced from fatty acids.

Lignans. Plant chemicals that act as antioxidants.

Linoleic acid. An omega-6 fatty acid found in vegetable oils, margarine, and processed foods.

Lipids. A family of chemical compounds that generally do not dissolve in water. Examples are fats and oils.

Lipid triad. A risk factor for heart disease characterized by the presence of elevated triglycerides (dietary fats not fully broken down by the liver that circulate in the blood), too-low HDL cholesterol, and high LDL cholesterol—in particular, a type of LDL cholesterol characterized by its small particle size.

Lipoprotein Lipase (LPL). An enzyme governing fat storage.

Lipoproteins. Protein blankets that carry fats to their destinations in the body.

Low-Density Lipoproteins (LDL). A type of cholesterol in the blood. High levels contribute to coronary heart disease.

Maltose. A simple sugar found in plants during the early stages of germination.

MCT oil (medium-chain triglyceride oil). A dietary fat metabolized in such a way that very little is stored as body fat.

Metabolic Rate. The speed at which your body burns calories.

Metabolic Syndrome (Syndrome X). A cluster of symptoms that set the stage for type II diabetes and heart disease. These symptoms include glucose intolerance, central obesity, elevated blood fats, and high blood pressure.

Metabolism. The physiological process that converts food to energy so the body can function.

Micelle. A tiny droplet made of lipids and emulsifiers from bile. They package cholesterol for transport in the body.

Mineral. A class of nutrients needed by the body for a wide range of enzymatic and metabolic functions.

Monosaccharide. A type of simple sugar constructed of a single molecule of glucose.

Monounsaturated Fat. Fatty acids that lack two hydrogen atoms. Found in such foods as olive oil, olives, avocados, cashew nuts, and cold-water fish such as salmon, mackerel, halibut, and swordfish.

Obesity. An excessive and abnormal amount of weight, usually 20 percent or more above a person's ideal weight. Obesity is a risk factor for heart disease, cancer, and type II diabetes.

Oleic acid. A monounsaturated fat found most notably in olive oil.

Oleuropein. The most abundant polyphenol, an antioxidantlike substance, found in olives and olive oil.

Olive oil. A healthy monounsaturated fat pressed from olives.

Omega-3 fatty acids. Essential fats found in fish and plants that appear to prevent blood clots and the buildup of plaque on arterial walls. Omega-3 fatty acids also play a role in strengthening the immune system.

Omega-6 fatty acids. Essential fats that are generally necessary for normal growth, hair and skin health, regulation of metabolism, and reproduction.

Oxidation. A chemical process in which oxygen combines with another substance, which is changed to another form.

Perilla oil. A supplement oil containing high amounts of alpha-linolenic acid.

Phospholipids. Fatlike substances that contain a molecule of phosphorus, which makes them soluble in water. This characteristic helps fats travel in and out of the lipid-rich membranes of cells.

Photosynthesis. A process in which the sun's energy is captured by chlorophyll, the green coloring in leaves, to turn water from the soil and carbon dioxide from the air into an energy-yielding sugar called glucose.

Phytochemicals. A large group of health-bestowing chemicals found in plant foods.

Plant sterols. Naturally occurring substances present in small quantities in many fruits, vegetables, legumes, nuts, seeds, grains, and other plant sources. Used in cholesterol-lowering margarines.

Platelets. Tiny clotting factors in blood.

Polyphenols. Disease-fighting substances that play roles in cardiovascular health and cancer protection.

Polysaccharide. Multiple number of sugar molecules linked together in a long chain. Complex carbohydrates (starches) are polysaccharides.

Polyunsaturated Fat. A fatty acid that lacks four or more hydrogen atoms. Found in fish and in most vegetable oils.

Prebotics. A type of undigested carbohydrate technically known as "fructooligosaccharides," or FOS, for short. FOS become a "meal" for health-promoting bacteria such as acidophilus (known as a probiotic), believed to enhance healthy flora in the intestines, improve digestion, and prevent disease.

Prostacyclin. Synthesized mostly from EPA, prostacyclin orders platelets to not stick together and move along.

Prostaglandins. Hormonelike compounds made from fatty acids that regulate nearly every system in the body.

Protein. A food group necessary for growth and repair of body tissues.

Ribose. A beneficial simple sugar that forms the carbohydrate backbone of DNA and RNA, the genetic material that controls cellular growth and reproduction, thus governing all life. Available as a supplement marketed to exercisers and athletes.

Saturated Fat. A fatty acid that is solid at room temperature.

Saturation. The number of hydrogens in a fatty acid chain.

Sesame oil. A cooking oil extracted from sesame seeds. It contains a powerful antioxidant called sesaminol.

Simple Sugars. A type of carbohydrate that is constructed of either single or double molecules of glucose.

Soluble Fiber. A type of fiber in grains, legumes, and carrots that has been shown to reduce cholesterol and slow the release of glucose into the bloodstream.

Squalene. A vitaminlike chemical found in olives, olive oil, and shark liver oil.

Starch. A type of carbohydrate consisting of three or more glucose molecules. Most plant foods, including cereals, whole grains, pasta, fruits, and vegetables, are complex carbohydrates.

Stearic acid. A saturated fat that does not cause heart problems.

Stearidonic acid. A by-product of alpha-linolenic acid that may be responsible for many of the health benefits associated with black currant seed oil.

Strength Training. Any kind of weight-bearing activity in which your muscles are challenged to work harder each time they're exercised.

Stroke. Bleeding or lack of blood supply in the brain.

Sugar. A form of carbohydrate that supplies calories and can elevate blood glucose levels.

Sugar Substitutes. Sweeteners that can be used in place of sugar to reduce calories or control glucose levels. Substitutes include saccharin, aspartame, acesulfame-K, sugar alcohols, stevia (an herb), and tagatose.

Tagatose. A low-calorie natural sugar that has been recently approved by the FDA for use in foods, beverages, and other products.

Thromboxane. A prostaglandin mainly synthesized from arachidonic acid that tells platelets—tiny clotting factors in your blood—to clump together.

Trans-fatty acids. By-products of hydrogenation that are responsible for causing serious diseases.

Triglycerides. Fats that circulate in the blood until they are deposited in fat cells. Elevated triglycerides (above 200) are a risk factor for heart disease.

Tropical oils. A group of vegetable fats that are saturated. They include coconut oil, palm and palm kernel oils, and the cocoa butter in chocolate.

Type I Diabetes. A form of diabetes that develops prior to age thirty but can occur at any age. It is caused by an immune system attack on the body's own beta cells. When these cells are destroyed, insulin can no longer be produced.

Type II Diabetes. A form of diabetes that occurs in people over age forty but can develop in younger people. With type II diabetes, the body does not use insulin properly.

Tyrosol. A major polyphenol in olives and olive oil. It functions as an antioxidant to help prevent the oxidation of LDL cholesterol and keep free radicals from attacking cell membranes.

Unsaturated Fat. Fats that are liquid at room temperature.

Very-low-density lipoproteins. A harmful type of cholesterol.

Vitamins. Organic substances found in food that perform many vital functions in the body.

Wheat germ oil. Extracted from wheat germ and sold as a nutritional supplement, wheat germ oil contains a number of healthful substances, including octacosanol, an alcohol derivative that may be heart-healthy; vitamin E, an important antioxidant; and several minerals.

•REFERENCES•

Good Carbs vs. Bad Carbs

A portion of the information in this book comes from medical research reports in both popular and scientific publications, professional textbooks and booklets, books, Internet sources, and computer searches of medical databases of research abstracts.

Chapter 1: The Food Fuel

Brown, J. *The Science of Human Nutrition*. San Diego: Harcourt Brace Jovanovich, 1990.

Deutsch, R. M., *et al. Realities of Nutrition*. Palo Alto, CA: Bull Publishing Company, 1993.

Herbert, V., *et al. Total Nutrition*. New York: St. Martin's Press, 1995.

Kleiner, S. M. "Antioxidant Answers." *The Physician and Sportsmedicine*. 24:21–22, 1996.

Jequier, E. "Carbohydrates as a Source of Energy." *American Journal of Clinical Nutrition*. 59:682S–685S, 1994.

Chapter 2: Carbohydrate Scorecards

Begley, S. "Beyond Vitamins." *Newsweek,* April 25, 1994, pp. 44–49.

Blumenthal, D. "A Simple Guide to Complex Carbs." *FDA Consumer.* 23:13–17, 1989.

"Diet. The Glycemic Index." *Harvard Health Letter.* 24:7, 1999.

Foster-Powell, K., *et al.* "International Tables of Glycemic Index." *American Journal of Clinical Nutrition.* 62:871S–890S, 1995.

Frost, G. "The Relevance of the Glycemic Index to Our Understanding of Dietary Carbohydrate." *Diabetic Medicine.* 17:336–345, 2000.

Glycemic Index Research Institute. *The Complete Guide to Fat-Storing Carbohydrates.* Washington, DC: Glycemic Index Research Institute, 2000.

Kleiner, S. M. "Antioxidant Answers." *The Physician and Sportsmedicine.* 24:21–22, 1996.

Morris, K. L., *et al.* "Glycemic Index, Cardiovascular Disease, and Obesity." *Nutrition Review.* 57:273–276, 1999.

Chapter 3: Healthy Bites: The Twenty-five Super Carbs

"Gut Guard: Vitamin E Against Tumor-Friendly Damage." *Prevention*, February 1994, pp. 18–19.

Gebhardt, R. "Antioxidant and Protective Properties of Extracts from Leaves of the Artichoke (*Cynara scolymus L.*) Against Hydroperoxide-Induced Oxidative Stress in Cultured Rat Hepatocytes." *Toxicology and Applied Pharmacology.* 144:279–286, 1997.

McCord, H. "Nature's Best Cholesterol Crunchers." *Prevention*, May 1, 1994, pp. 54–62.

Ness, A. R., *et al.* "Fruit and Vegetables, and Cardiovascular Disease: A Review." *International Journal of Epidemiology.* 26:1–13, 1997.

Segasothy, M., *et al.* "Vegetarian Diet: Panacea for Modern Lifestyle Diseases?" *Quarterly Journal of Medicine.* 92:531–544, 1999.

"Vitamins, Carotenoids, and Phytochemicals." www.webmd.com, 1999.

Chapter 4: Your Brain on Carbs

Blaun, R. "How to Eat Smart." *Psychology Today.* 29:34–44, 1996.

Greenwood-Robinson, M. *20/20 Thinking.* New York: Avery Books, 2003.

Jeerkathil, T. J., *et al.* "Prevention of Strokes." *Current Atherosclerosis Reports.* 3:321–327, 2001.

Korol, D. L., *et al.* "Glucose, Memory, and Aging." *American Journal of Clinical Nutrition.* 67:746S–671S, 1998.

Trankina, M. "Choosing Foods to Modulate Moods." *The World & I.* 13:150, 1998.

Wurtman, J. J. "Carbohydrate Cravings: A Disorder of Food Intake and Mood." *Clinical Neuropharmacology.* 11:S139–145, 1988.

Wurtman, J. J. "Effect of Nutrient Intake on Premenstrual Depression." *American Journal of Obstetrics and Gynecology.* 161:1228–1234, 1989.

Wurtman, J. J. "Nutrients Affecting Brain Composition and Behavior." *Integrative Psychiatry.* 5:226–238, 1987.

Chapter 5: The Carb-Cancer Connection

American Institute for Cancer Research. "Food, Nutrition, and the Prevention of Cancer: A Global Perspective." www.aicr.org, 2002.

———. "Simple Steps to Prevent Cancer." www.aicr.org, 2002.

Associated Press. "Cancer Could Rise 50 Percent, U.S. Says." *Daily News.* April 4, 2003, p. B6.

Augustin, L. S., *et al.* "Dietary Glycemic Index and Glycemic Load, and Breast Cancer Risk: A Case-Controlled Study." *Annals of Oncology.* 12:1533–1538, 2001.

Berrino, F., *et al.* "Reducing Bioavailable Sex Hormones Through a Comprehensive Change in Diet: The Diet and Androgens (DIANA) Randomized Trial." *Cancer Epidemiology, Biomarkers & Prevention.* 10:25–33, 2001.

De Stefani, E., *et al.* "Dietary Sugar and Lung Cancer: A Case-Controlled Study in Uruguay." *Nutrition and Cancer.* 31:132–137, 1998.

Fox, M. C. "Control Carbs, Control Cancer." *Prevention.* January 1, 2003, p. 119.

Franceschi, S., *et al.* "Dietary Glycemic Load and Colorectal Cancer Risk." *Annals of Oncology.* 12:173–178, 2001.

Michaud, D. S., *et al.* "Dietary Sugar, Glycemic Load, and Pancreatic Cancer Risk in a Prospective Study." *Journal of the National Cancer Institute.* 94:1293–1300, 2002.

Nijveldt, R. J. "Flavonoids: A Review of Probable Mechanisms of Action and Potential Applications." *American Journal of Clinical Nutrition.* 74:418–425, 2001.

Slattery, M. I., *et al.* "Dietary Sugar and Colon Cancer." *Cancer Epidemiology, Biomarkers & Prevention.* 6:677–685, 1997.

Slavin, J. I., *et al.* "Plausible Mechanisms for the Protectiveness of Whole Grains." *American Journal of Clinical Nutrition.* 70:459S–463S, 1999.

Steinmetz, K. A., *et al.* "Vegetables, Fruit, and Cancer Prevention: A Review." *Journal of the American Dietetic Association.* 96:1027–1039, 1996.

Tipton, M., "Selenium May Cut the Risk of Prostate Cancer." *Medical Update.* 22:3, 1998.

Chapter 6: The Diabetes Defense

American Diabetes Association. "Diabetes Info." www. diabetes.org, 2001.

———. "Evidence-Based Nutrition Principles and Recommendations for the Treatment and Prevention of Diabetes and Related Complications." *Diabetes Care.* 26:S51–S61, 2003.

———. "Nutrition Guide for People with Diabetes." www.diabetes.org, 2003.

Arcement, P. S. "Carbohydrate Counting in Diabetes Meal Planning." *Home Healthcare Nurse.* 17:425–428, 1999.

Benedict, M. "Carbohydrate Counting. Tips for Simplifying Diabetes Eduction." *Health Care Food & Nutrition Focus.* 16:6–9, 1999.

Daly, A., *et al.* "Carbohydrate Counting: Getting Started." Chicago, IL: American Dietetics Association, 1995.

Greenwood-Robinson, M. *Control Diabetes in 6 Easy Steps.* New York: St. Martin's Press, 2002.

Chapter 7: Digestive Health

Black, J. "Acute Appendicitis in Japanese Soldiers in Burma: Support for the "Fibre" Theory." *Gut.* 51:297, 2002.

Brown, E. W. "50,000 Unnecessary Appendectomies a Year—and How to Prevent Them." *Medical Update.* 20:4, 1997.

Carson-Dewitt, R. S. "Diverticulosis and Diverticulitis." *Gale Encyclopedia of Medicine.* www.findarticles.com, 1999.

Dosh, S. A. "Evaluation and Treatment of Constipation." *Journal of Family Practice*. www.findarticles.com, June 2002.

Flagg, S., *et al*. "Gut Protection." *Prevention*. January 1, 1997, pp. 31–32.

Haggerty, M. "Constipation." *Gale Encyclopedia of Medicine*. www.find articles.com, 1999.

Marlatt, J., *et al*. "Position of the American Dietetic Association: Health Implications of Dietary Fiber." *Journal of the American Dietetic Association*. 102:993–1000, 2002.

Murray, M. T. *Encyclopedia of Nutritional Supplements*. Rocklin, Cal.: Prima Publishing, 1996.

Ortega, R. M., *et al*. "Differences in Diet and Food Habits Between Patients with Gallstones and Controls." *Journal of the American College of Nutrition*. 16:88–95, 1997.

Paradox. P. "Gallstones." *Gale Encyclopedia of Alternative Medicine*. www.findarticles.com, 2001.

Swade, S. "Ease Gut Reactions." *Prevention*. February 1, 1995, pp. 78–80.

Todoroki I., *et al*. "Cholecystectomy and the Risk of Colon Cancer." *American Journal of Gastroenterology*. 94:41–46, 1999.

Trautwein, E. A. "Dietetic Influences on the Formation and Prevention of Cholesterol Gallstones." *Zeitschrift für Ernahrungswissenschaft*. 33:2–15, 1994.

Walker, A. R., *et al*. "What Causes Appendicitis?" *Journal Clinical Gastroenterology*. 12:127–129, 1990.

Weil, A. "Warning: 11 Medical Practices to Avoid." *East West Natural Health*. 22:58–63, 1992.

Wolnik, L., and H. Bauer. "Epidemiology of Ovarian Cancer." *Onkologie*. 2:96–101, 1979.

Chapter 8: Heart-Healthy Carbs

Fallon, S., *et al*. "Diet and Heart Disease: Not What You Think." *Consumers' Research Magazine*. July 1, 1996, pp. 15–19.

Gee, S. "The AGEing Process (Advanced Glycation End Products)." *Diabetes Forecast*. 51:72–74, 1998.

Howard, B. V., *et al*. "Sugar and Cardiovascular Disease." *Circulation*. 106:523–527, 2002.

Hunter, B. T. "Confusing Consumers About Sugar Intake." *Consumers' Research Magazine*. January 1, 1995, pp. 14–17.

"Inflammation and Heart Disease." *Medical Update.* January 1, 2002, p. 6.

Katan, M. B. "Are There Good and Bad Carbohydrates for HDL Cholesterol?" *The Lancet.* 353:1029–1030, 1999.

Lemonick, M. "Lean and Hungrier Is a Recently Discovered Hormone." *Time.* June 3, 2002. p. 54.

Ness, A. R., *et al.* "Fruit and Vegetables, and Cardiovascular Disease: A Review." *International Journal of Epidemiology.* 26:1–13, 1997.

Park, A. "Beyond Cholesterol: Inflammation Is Emerging as a Major Risk Factor—and Not Just in Heart Disease." *Time.* November 25, 2002, p. 75.

Squires, S. "Sweet but Not So Innocent." *The Washington Post.* March 11, 2003, p. F01.

Chapter 9: Carbs and Weight Control:
The Good-Carb Diet

Ammon, P. K. "Individualizing the Approach to Treating Obesity." *Nurse Practitioner.* 24:27–31, 36–38, 1999.

Auchmutey, J. "Sugar Nation." *The Atlanta Journal-Constitution.* November 17, 2002, p. A1.

Burton-Freeman, B. "Dietary Fiber and Energy Regulation." *Journal of Nutrition.* 130:272S–275S, 2002.

Howarth, N. C., *et al.* "Dietary Fiber and Weight Regulation." *Nutrition Reviews.* 59:129–139, 2001.

Kersten, S. "Mechanisms of Nutritional and Hormonal Regulation of Lipogenesis." *EMBO Reports.* 282–286, 2001.

Ludwig, D. S. "Dietary Glycemic Index and Obesity." *Journal of Nutrition.* 130:280S–283S, 2000.

Roberts, S. B., *et al.* "The Influence of Dietary Composition on Energy Intake and Body Weight." *Journal of the American College of Nutrition.* 21:140S–145S, 2002.

Chapter 10: Carb Power for Exercisers and Athletes

Akermark, C., *et al.* "Diet and Muscle Glycogen Concentration in Relation to Physical Performance in Swedish Elite Ice Hockey Players." *International Journal of Sports Nutrition.* 6:272–284, 1996.

Burke, L. M. "Nutrition for Post-Exercise Recovery." *International Journal of Sports Nutrition.* 1:214–224, 1997.

Chandler R. M., *et al.* "Dietary Supplements Affect the Anabolic Hormones After Weight-Training Exercise." *Journal of Applied Physiology.* 76: 839–45, 1994.

Coyle, E. F. "Timing and Method of Increased Carbohydrate Intake to Cope with Heavy Training, Competition and Recovery." *Journal of Sports Sciences.* 9:29–51, 1991.

"How Do Sports Drinks Work?" Barrington, IL: The Gatorade Company, 2000.

Ivy, J. L. "Glycogen Resynthesis After Exercise: Effect of Carbohydrate Intake." *International Journal of Sports Medicine.* 19:S142–S145, 1998.

———. "Role of Carbohydrate in Physical Activity." *Clinics in Sports Medicine.* 18:469–484, 1999.

Kleiner, S. *Power Eating.* Champaign, IL: Human Kinetics, 1998.

Chapter 11: Your Smart-Carb Strategy

American Cancer Society. "ACS Nutrition Guidelines." www.cancer.org, 2003.

Good Fat vs. Bad Fat

A portion of the information in this book comes from medical research reports in both popular and scientific publications, professional textbooks and booklets, promotional literature from supplement companies, case studies, Internet sources, and computer searches of medical databases of research abstracts.

Chapter 1: The Fats of Life

Carper, J. 1995. *Stop aging now!* New York: HarperPerennial.

Editor. 2001. Frequently asked questions. Online: www. fatsforhealth.com.

Herbert, V. (ed.). 1995. *Total nutrition: The only guide you'll ever need.* New York: St. Martin's Press.

Kettler, D. B. 2001. Can manipulation of the ratios of essential fatty acids slow the rapid rate of postmenopausal bone loss? *Alternative Medicine Review* 6: 61–77.

Lawton, C. L., et al. 2000. The degree of saturation of fatty acids influences post-ingestive satiety. *British Journal of Nutrition* 83: 473–482.

Murray, M. T. 1996. *Encyclopedia of nutritional supplements.* Rocklin, California: Prima Publishing.

Sizer, F., and E. Whitney. 1997. *Nutrition concepts and controversies.* 7th ed. Belmont, California: West/Wadsworth.

Chapter 2: When Fat Can Be Fatal

Applegate, L. 1994. Fat transformed. *Runner's World,* February, pp. 26–27.

Carper, J. 1995. *Stop aging now!* New York: HarperPerennial.

Chong, Y. H., et al. 1991. Effects of palm oil on cardiovascular risk. *The Medical Journal of Malaysia* 46: 41–50.

Collins, S. 2001. Your health: Chocolate is good for you. *Sunday Mirror,* April 15, pp. 44–45.

Connor, W. E. 1999. Harbingers of coronary heart disease: dietary saturated fatty acids and cholesterol. Is chocolate benign because of its stearic acid content? *American Journal of Clinical Nutrition* 70: 951–952.

De La Taille, A., et al. 2001. Cancer of the prostate: influence of nutritional factors. General nutritional factors. *Presse Medicale* 30: 554–556.

Ebong, P. E., et al. 1999. Influence of palm oil (Elaesis guineensis) on health. *Plant Foods for Human Nutrition* 53: 209–222.

Editor. 1996. The good news about "bad" foods. *Good Housekeeping,* September, pp. 93–94.

Editor. 1992. Relax, Mrs. Sprat high-fat, low-fiber diets may not cause breast cancer after all. *Time,* November 2, p. 23.

Editor. 1998. Researchers sweet on health benefits of chocolate (stearic acid). *Medical Post,* September 22, p. 59.

Elson, C. E. 1992. Tropical oils: nutritional and scientific issues. *Critical Reviews in Food and Science Nutrition* 31: 79–102.

Eritson, J. 2000. Safety considerations of polyunsaturated fatty acids. *American Journal of Clinical Nutrition* 71: 197–201.

Fallon, S., et al. 1996. Diet and heart disease: not what you think. *Consumers' Research Magazine,* July, pp. 15–19.

Fallon, S. 1996. Why butter is good for you. *Consumers' Research Magazine,* March, pp. 10–15.

Herbert, V. (ed.). 1995. *Total nutrition: The only guide you'll ever need.* New York: St. Martin's Press.

Hooper, J. 1998. What you still don't know about cholesterol. *Esquire,* March, pp. 136–138.

Lichtenstein, A. H., et al. Dietary fat consumption and health. *Nutrition Reviews* 56: S3–S28.

Murray, M. T. 1996. *Encyclopedia of nutritional supplements*. Rocklin, California: Prima Publishing.

Nash, J. M. 1994. Is a low-fat diet risky? *Time,* September 5, p. 62.

Sadler, C. 1998. The fat came back. *Chatelaine,* April, pp. 163–164.

Sizer, F., and E. Whitney. 1997. *Nutrition concepts and controversies*. 7th ed. Belmont, California: West/Wadsworth.

Stampfer, M. J., et al. 2000. Primary prevention of coronary heart disease in women through diet and lifestyle. *New England Journal of Medicine* 343: 16–22.

Wells, A. S., et al. 1998. Alterations in mood after changing to a low-fat diet. *British Journal of Nutrition* 79: 23–30.

Chapter 3: Fishing for Good Health

Appel, A. 1999. A fish tale. *Natural Health,* April. Online: www.findarti cles.com.

Appel, L. J., et al. 1993. Does supplementation of diet with fish oil reduce blood pressure? A meta-analysis of controlled clinical trials. *Archives of Internal Medicine* 153: 1429–1438.

Applegate, L. 1999. Make room for fish. *Runner's World,* October, pp. 26–28.

Belluzzi, A., et al. 2000. Polyunsaturated fatty acids and inflammatory bowel disease. *American Journal of Clinical Nutrition* 71: 339S–342S.

Brewer, S. 2001. Family health. *Daily Record,* February 28, p. 6.

Bucci, L. 1993. Nutrients as ergogenic aids for sports and exercise. Boca Raton, Florida: CRC Press.

Carper, J. 1995. *Stop aging now*! New York: HarperPerennial.

Connor, W. E., et al. 1993. N-3 fatty acids from fish oil. *Annals of the New York Academy of Sciences* 683: 16–34.

Durham, J. 1999. The skinny on fat. *Saturday Evening Post,* May.

Editor. 1997. Schizophrenia symptoms eased with fish oil. *Medical Post,* September 9, p. 61.

Freeman, M. P. 2000. Omega-3 fatty acids in psychiatry: a review. *Annals of Clinical Psychiatry* 12: 159–165.

Goodman, J. 2001. *The omega solution*. Rocklin, California: Prima Publishing.

Gorman, C., et al. 2000. They're full of a special fat called omega-3 that may actually be good for you. *Time,* October 30, p. 76.

Halpern, G. M. 2000. Anti-inflammatory effects of a stabilized lipid extract of Perna canaliculus (Lyprinol). *Allergie et Immunologie* 32: 272–278.

Herbert, V. (ed.). 1995. *Total nutrition: The only guide you'll ever need.* New York: St. Martin's Press.

Jobrin, J. 1996. Tackle arthritis with a knife and fork. *Prevention,* November, pp. 83–91.

Joy, C. B., et al. 2000. Polyunsaturated fatty fish (fish or evening primrose oil) for schizophrenia. Cochrane Database System Review.

Kremer, J. M., et al. 1995. Effects of high-dose fish oil on rheumatoid arthritis after stopping nonsteroidal antiinflammatory drugs. Clinical and immune correlates. *Arthritis and Rheumatism* 38: 1107–1114.

Leaf, A. 1992. The role of omega-3 fatty acids in the prevention and rehabilitation of coronary artery disease. *Annals of the Academy of Medicine* 21: 132–136.

McCord, H. 1999. Nutrition news: this oil's a lifesaver. *Prevention,* October, p. 56.

McCord, H. 1997. The fat you need. *Prevention,* January, pp. 100–108.

Nettleton, J. A. 1991. Omega-3 fatty acids: comparison of plant and seafood sources. *Journal of the American Dietetic Association* 91: 331–337.

Pugliese, P. T., et al. 1998. Some biological actions of alkylglycerols from shark liver oil. *Journal of Alternative and Complementary Medicine* 4: 87–99.

SerVaas, C., et al. 1999. Fats for mental health. *Saturday Evening Post,* March.

Simopoulos, A. P. 1999. Evolutionary aspects of omega-3 fatty acids in the food supply. *Prostaglandins, Leukotrienes, and Essential Fatty Acids* 60: 421–429.

Sinclair, A. J., et al. 2000. Marine lipids: overview "new insights and lipid composition of Lyprinol." *Allergie et Immunologie* 32: 261–271.

Webb, D. 1999. Healthy diet: the smart fat makeover. *Prevention,* April, pp. 134–141.

Webb, D. 2000. Can fish oil chase away both heart attacks and blues? *Prevention,* March, pp. 69–70.

Yetiv, J. Z. 1988. Clinical applications of fish oil. *Journal of the American Medical Association* 260: 665–670.

Chapter 4: DHA: The Brain-Building Fat

Brody, J. 2001. Science sees new diet role in eye care. *Minneapolis Star Tribune,* March 21, p. 04E.

Cooper, R. 1998. *DHA: the essential omega-3 fatty acid.* Pleasant Grove, Utah.

Dickstein, L. 1999. DHA: the good fat. *Psychology Today,* April, p. 50.

Editor. 1997. Alzheimer's, depression, attention-deficit/hyperactivity disorder linked to low levels of DHA, a key brain fat. April 9. Online: www.royal-health.com.

Editor. 2001. DHA: the mind mender. *Psychology Today,* March, p. 48.

Editor. 1997. Lack of breast milk may be risk factor in schizophrenia. May 15. Online: www.royal-health.com.

Editor. 1995. Two studies report Martek's DHA—Neuromins™—improves HDL:LDL cholesterol ration and lowers triglycerides. July 11. Online: www.royal-health.com.

Gamoh, S., et al. 1999. Chronic administration of docosahexaenoic acid improves reference memory–related learning ability in young rats. *Neuroscience* 93: 237–241.

Horrocks, L. A., et al. 1999. Health benefits of docosahexaenoic acid (DHA). *Pharmacological Research* 40: 211–225.

Mori, T. A., et al. 1999. Docosahexaenoic acid but not eicosapentaenoic acid lowers ambulatory blood pressure and heart rate in humans. *Hypertension* 34: 253–260.

Murray, M. T. 1999. Must-have nutrients for mothers-to-be. *Better Nutrition,* August. Online: www.findarticles.com.

Shimura, T., et al. 1997. Docosahexaenoic acid (DHA) improved glucose and lipid metabolism in KK-Ay mice with genetic non-insulin-dependent diabetes mellitus (NIDDM). *Biological & Pharmaceutical Bulletin* 20: 507–510.

Waltman, A. B., et al. 2000. Guide to natural health: alternative medicine goes mainstream. *Psychology Today,* April, pp. 37–40, 42.

Chapter 5: Nature's Disease Fighters

Allman, M. A., et al. 1995. Supplementation with flaxseed oil versus sunflower seed oil in healthy young men consuming a low fat diet: effects on platelet composition and function. *European Journal of Clinical Nutrition* 49: 169–178.

American Lung Association. 2001. Acute respiratory distress syndrome (ARDS). Online: www.lungusa.org.

Annuseek, G. 2001. Flaxseed. *Gale Encyclopedia of Alternative Medicine*. Online: www.findarticles.com.

Blumenthal, M. 1998. *The complete German Commission E monographs: therapeutic guide to herbal medicines*. American Botanical Council: Austin, Texas.

Craig, W. J. 1999. Health-promoting properties of common herbs. *American Journal of Clinical Nutrition* 70: 491S–499S.

Cunnane, S. C., et al. 1993. High alpha-linolenic acid flaxseed (linum usitatissimum): some nutritional properties in humans. *British Journal of Nutrition* 69: 443–453.

Editor. 1999. Foods that fight breast cancer. *Saturday Evening Post,* January.

Engler, M. M., et al. 1998. Dietary borage oil alters plasma, hepatic and vascular tissue fatty acid composition in spontaneously hypertensive rats. *Prostaglandins, Leukotrienes, and Essential Fatty Acids* 59: 11–15.

Engler, M. M., et al. Dietary gamma-linolenic acid lowers blood pressure and alters aortic reactivity and cholesterol metabolism in hypertension. *Journal of Hypertension* 10: 1197–1204.

Ezaki, O., et al. 1999. Long-term effects of dietary alpha-linolenic acid from perilla oil on serum fatty acids composition and on the risk factors of coronary heart disease in Japanese elderly subjects. *Journal of Nutritional Science and Vitaminology* 45: 759–772.

Flax Council of Canada. 2001. *Flaxseed*. Winnipeg, Canada: Flax Council of Canada.

Grant, S. 1996. The beauty of borage. December 5, *The Evening Post,* p. 23.

Hartman, I. S. 1995. Alpha-linolenic acid: a prevention in secondary coronary events? *Nutrition Reviews* 53: 194–197.

Jacob, J., et al. 2000. Designer and specialty eggs. The Institute of Food and Agricultural Sciences, University of Florida.

James, M. J., et al. 2000. Dietary polyunsaturated fatty acids and inflammatory mediator production. *American Journal of Clinical Nutrition* 71: 343S–348S.

Jenab, M., et al. 1996. The influence of flaxseed and lignans on colon carcinogenesis and beta-glucuronidase activity. *Carcinogenesis* 17: 1343–1348.

Leventhal, L. J., et al. Treatment of rheumatoid arthritis with gammalinolenic acid. *Annals of Internal Medicine* 119: 867–873.

Masataka, O., et al. 1997. Perilla oil prevents the excessive growth of visceral adipose tissue in rats by down-regulating adipocyte differentiation. *Journal of Nutrition* 127: 1752–1757.

Medical Economics Company. 2001. *The PDR for Herbal Medicines.* Montvale, New Jersey: Medical Economics Company. Online: www.pdr.net.

Miller, L. G. 1998. Herbal medicinals: selected clinical considerations focusing on known or potential drug-herb interactions. *Archives of Internal Medicine* 158: 2200–2211.

Mills, D. E., et al. 1989. Dietary fatty acid supplementation alters stress reactivity and performance in man. *Journal of Human Hypertension* 3: 111–116.

Murray, M. T. 1996. *Encyclopedia of nutritional supplements.* Rocklin, California: Prima Publishing.

Narisawa, T., et al. 1994. Colon cancer prevention with a small amount of dietary perilla oil high in alpha-linolenic acid in an animal model. *Cancer* 73: 2069–2075.

Okamoto, M., et al. 2000. Effects of dietary supplementation with n-3 fatty acids compared with n-6 fatty acids on bronchial asthma. *Internal Medicine* 39: 107–111.

Prasad, K. 1997. Dietary flax seed in prevention of hypercholesterolemic atherosclerosis. *Atherosclerosis* 132: 69–76.

Prasad, K. 2000. Oxidative stress as a mechanism of diabetes in diabetic BB prone rats: effect of secoisolariciresinol diglucoside (SDG). *Molecular and Cellular Biochemistry* 209: 89–96.

Prasad, K. 1999. Reduction of serum cholesterol and hypercholesterolemic atherosclerosis in rabbits by secoisolariciresinol diglucoside isolated from flaxseed. *Circulation* 99: 1355–1362.

Shahidi, F. 2000. Antioxidant factors in plant foods and selected oilseeds. *Biofactors* 13: 179–185.

Sung, M. K., et al. 1998. Mammalian lignans inhibit the growth of estrogen-independent human colon tumor cells. *Anticancer Research* 18: 1405–1408.

Swift, D. 1999. Natural oil improves odds of ARDS recovery. *Medical Post,* November 17.

Thompson, L. U., et al. 1996. Flaxseed and its lignan and oil components reduce mammary tumor growth at a late stage of *Carcinogenesis* 17: 1373–1376.

Udall, K. G. 1997. *Flaxseed oil*. Pleasant Grove, Utah: Woodland Publishing.

Yan, L., et al. 1998. Dietary flaxseed supplementation and experimental metastasis of melanoma cells in mice. *Cancer Letter* 124: 181–186.

Chapter 6: Omega-6 Healers

Ahn, Y. O. 1997. Diet and stomach cancer in Korea. *International Journal of Cancer* 10: S7–S9.

Blumenthal, M. 1998. *The complete German Commission E monographs: therapeutic guide to herbal medicines*. American Botanical Council: Austin, Texas.

Campbell, E. M., et al. 1997. Premenstrual symptoms in general practice patients. Prevalence and treatment. *Journal of Reproductive Medicine* 42: 637–646.

Cheung, K. L. 1999. Management of cyclical mastalgia in oriental women. *The Australian New Zealand Journal of Surgery* 69: 492–494.

Craig, W. J. 1999. Health-promoting properties of common herbs. *American Journal of Clinical Nutrition* 70: 491S–499S.

Darlington, L. G., et al. 2001. Antioxidants and fatty acids in the amelioration of rheumatoid arthritis and related disorders. *British Journal of Nutrition* 85: 251–269.

Editor. 2001. Consumer guide to wheat germ oil. Online: www.mother nature.com.

Editor. 1998. Hemp without the high. *Soap Perfumery & Cosmetics* 71: 34.

Editor. 1996. How to relieve vaginal dryness. *Meno Times,* June, p. 4.

Gateley, C. A., et al. 1992. Drug treatments for mastalgia: 17 years experience in the Cardiff Mastalgia Clinic. *Journal of the Royal Society of Medicine* 85: 12–15.

Gateley, C. A., et al. 1991. Management of the painful and modular breast. *British Medical Bulletin* 47: 284–294.

Hardy, M. L. 2000. Herbs of special interest to women. *Journal of the American Pharmaceutical Association* 40: 234–242.

Hilz, M. J., et al. 2000. Diabetic somatic polyneuropathy. *Fortschritte der Neurologie-Psychiatrie* 68: 278–288.

Jill, J. F., et al. 2000. Evening primrose oil and borage oil in rheumatologic conditions. *American Journal of Clinical Nutrition* 71: 352S–356S.

Kagi, M. K., et al. 1993. Falafel burger anaphylaxis due to sesame seed allergy. *Annals of Allergy* 71: 127–129.

Kang, M. H., et al. 2000. Mode of action of sesame lignans in protecting low-density lipoprotein against oxidative damage in vitro. *Life Sciences* 66: 161–171.

Kerscher, M. J., et al. 1992. Treatment of atopic eczema with evening primrose oil: rationale and clinical results. *The Clinical Investigator* 70: 167–171.

Miller, L. G. 1998. Herbal medicinals: selected clinical considerations focusing on known or potential drug-herb interactions. *Archives of Internal Medicine* 158: 2200–2211.

Murray, M. T. 1996. *Encyclopedia of nutritional supplements*. Rocklin, California: Prima Publishing.

Pritchard, G. A., et al. 1989. Lipids in breast carcinogenesis. *British Journal of Surgery* 76: 1069–1073.

Ramesh, G., et al. 1992. Effect of essential fatty acids on tumor cells. *Nutrition* 8: 343–347.

Salerno, J. W., et al. 1991. The use of sesame seed oil and other vegetable oils in the inhibition of human colon cancer growth in vitro. *Anticancer Research* 11: 209–215.

Smith, D. E., et al. 1992. Selective growth inhibition of a human malignant melanoma cell line by sesame in vitro. *Prostaglandins, Leukotrienes, and Essential Fatty Acids* 46: 145–150.

Struempler, R. E. 1997. A positive cannabinoids workplace drug test following the ingestion of commercially available hemp seed oil. *Journal of Analytical Toxicology* 21: 283–285.

Wagner, W., et al. 1997. Prophylactic treatment of migraine with gamma-linolenic and alpha-linolenic acids. *Cephalalgia* 17: 127–130.

Wu, D., et al. 1999. Effect of dietary supplementation with black currant seed oil on the immune response of healthy elderly subjects. *American Journal of Clinical Nutrition* 70: 536–543.

Chapter 7: Olive Oil: The Master Monounsaturated Fat

Ballmer, P. E. 2000. The Mediterranean diet—healthy but still delicious. *Therapeutische Umschau* 57: 167–172.

Bisignano, G., et al. 1999. On the in-vitro antimicrobial activity of oleuropein and hydroxytyrosol. *The Journal of Pharmacy and Pharmacology* 51: 971–974.

California Olive Industry. 2001. Information packet. Fresno, California: California Olive Industry.

California Olive Oil Council. 2000. Health news. Online: www.cooc.com.

Chan, P., et al. 1996. Effectiveness and safety of low-dose pravastatin and squalene, alone and in combination, in elderly patients with hypercholesterolemia. *Journal of Clinical Pharmacology* 36: 422–427.

Covas, M. I., et al. Virgin olive oil phenolic compounds: binding to human low density lipoprotein (LDL) and effect on LDL oxidation. *International Journal of Clinical Pharmacology Research* 20: 49–54.

Dupont, J., et al. 1989. Food safety and health effects of canola oil. *Journal of the American College of Nutrition* 8: 360–375.

Ferrara, L. A., et al. 2000. Olive oil and reduced need for hypertensive medications. *Archives of Internal Medicine* 160: 837–842.

Giovannini, C., et al. 1999. Tyrosol, the major olive oil biophenol, protects against oxidized-LDL-induced injury in caco-2 cells. *Journal of Nutrition* 129: 1269–1277.

Haban, P., et al. 2000. Oleic acid serum phospholipid content is linked with the serum total- and LDL-cholesterol in elderly subjects. *Medical Science Monitor* 6: 1093–1097.

Hammock, D., et al. 2001. Canola oil. May 30. Online: www.goodhouse keeping.women.com.

Horowitz, J. 1990. Food: a card game? No, cooking canola oil is the latest love of the cholesterol-free set. *Time,* November 12, p. 107.

Hunter, B. T. 1999. Modified vegetable oils. *Consumers' Research Magazine,* March.

Lee, A., et al. 2000. Consumption of tomato products with olive oil but not sunflower oil increases the antioxidant activity of plasma. *Free Radical Biology & Medicine* 29: 1051–1055.

Massaro, M., 1999. Direct vascular antiatherogenic effects of oleic acid: a clue to the cardioprotective effects of the Mediterranean diet. *Cardiologia* 44: 507–513.

McDonald, B. E. 2001. Canola oil: nutritional properties. Online: www. canola-council.org.

McGrath, M. 1999. How healthy is your olive oil? *Prevention,* September, pp. 122–127.

Newmark, H. L. 1997. Squalene, olive oil, and cancer risk: a review and hypothesis. *Cancer Epidemiology, Biomarkers & Prevention* 6: 1101–1103.

Nydahl, M., et al. 1995. Similar effects of rapeseed oil (canola oil) and olive oil in a lipid-lowering diet for patients with hyperlipoproteinemia. *Journal of the American College of Nutrition* 14: 643–651.

Ragionie, F. D., et al. 2000. Hyroxytyrosol, a natural molecule occurring in olive oil, induces cytochrome c-dependent apoptosis. *Biochemical and Biophysical Research Communications* 278: 733–739.

Ramirez-Tortosa, M. C., et al. 1999. Extra-virgin olive oil increases the resistance of LDL to oxidation more than refined olive oil in free-living men with peripheral vascular disease. *Journal of Nutrition* 129: 2177–2183.

Rodriguez-Villar, C., et al. 2000. High-monounsaturated fat, olive oil–rich diet has effects similar to a high-carbohydrate diet on fasting and postprandial state and metabolic profiles of patients with type 2 diabetes. *Metabolism* 49: 1511–1517.

Smith, T. J. 2000. Squalene: potential chemopreventive agent. *Expert Opinion on Investigational Drugs* 9: 1841–1848.

Trichopoulou, A., et al. 2000. Cancer and Mediterranean dietary traditions. *Cancer Epidemiology, Biomarkers & Prevention* 9: 869–873.

Visioli, F., et al. 1998. Free radical–scavenging properties of olive oil polyphenols. *Biochemical and Biophysical Research Communications* 9: 60–64.

Visioli, F., et al. 2000. Olive phenol hydroxytyrosol prevents passive smoking–induced oxidative stress. *Circulation* 102: 2169–2171.

Weil, A. 2000. What's the best olive oil? Ask Dr. Weil, June 6. Online: www.drweil.com.

Zambon, A., et al. 1999. Effects of hypocaloric dietary treatment enriched in oleic acid on LDL and HDL, subclass distribution in mildly obese women. *Journal of Internal Medicine* 246: 191–201.

Chapter 8: The New Heart-Saving Fats

American Dietetic Association. 1998. Fat replacers. Position statement. *Journal of the American Dietetic Association* 98: 463–468.

American Dietetic Association. 1995. Position of the American Dietetic Association: phytochemicals and functional foods. *Journal of the American Dietetic Association* 95: 493–496.

Blair, S. N. 2000. Incremental reduction of serum total cholesterol and low-density lipoprotein cholesterol with the addition of plant stanol ester-containing spread to statin therapy. *American Journal of Cardiology* 86: 46–52.

Editor. 2000. Can the new plant-based spreads really lower cholesterol? *John Hopkins Medical Letter Health After 50* 12: 98.

Gorman, C. 1999. It sure ain't butter. *Time,* May 31, p. 104.

Gylling, H., et al. 1997. Reduction of serum cholesterol in postmenopausal women with previous myocardial infarction and cholesterol malabsorption induced by dietary sitostanol ester margarine. *Circulation* 96: 4226–4231.

Hallikainen, M. A., et al. 1999. Effects of 2 low-fat stanol ester-containing margarines on serum cholesterol concentrations as part of a low-fat diet in hypercholesterolemic subjects. *American Journal of Clinical Nutrition* 69: 403–410.

Kulman, L. 1999. I really can believe it's not butter. *U.S. News & World Report,* May 31, p. 77.

Lipton. 2001. Take Control. Online: www.takecontrol.com.

McCord, H. 1999. Spread your bread. Help your heart. *Prevention,* October, p. 55.

McNeil Consumer Healthcare. 2001. Benecol. Online: www.benecol.com.

Miettinen, T. A., et al. 1995. Reduction of serum cholesterol with sitostanol-ester margarine in a mildly hypercholesterolemic population. *New England Journal of Medicine* 333: 1350–1351.

Plat, J., et al. 2000. Therapeutic potential of plat sterols and stanols. *Current Opinion in Lipidology* 11: 571–576.

Rosenthal, R. L. 2000. Effectiveness of altering serum cholesterol levels without drugs. *BUMC Proceedings* 13: 351–355.

Weststrate, J. A., et al. Plant sterol-enriched margarines and the reduction of plasma total and LDL cholesterol concentrations in normocholesterolaemic and mildly hypercholesterolaemic subjects. *European Journal of Clinical Nutrition* 52: 334–343.

Chapter 9: Conjugated Linoleic Acid: Health Insurance in a Pill

American Cancer Society. 2001. Statistics. Online: www.cancer.org.

American Heart Association. 2001. *2001 Heart and Stroke Statistical Update.* American Heart Association: Dallas, Texas, p. 4.

American Society of Bariatric Physicians. 2000. What is obesity? Online: www.asbp.org/obesity.htm.

Basu, S., et al. 2000. Conjugated linoleic acid induces lipid peroxidation in humans. *FEBS Letter* 468: 33–36.

Cesano, A., et al. 1998. Opposite effects of linoleic acid and conjugated linoleic acid on human prostatic cancer in SCID mice. *Anticancer Research* 18: 1429–1434.

Editor. 1996. Compounds in milk may reduce early indicators of cancer. *Cancer Weekly Plus,* November 11, 8.

Editor. 2001. Conjugated linoleic acid overview. Professional Monographs: Herbal, Mineral, Vitamin, Nutraceuticals. Intramedicine, March 1.

Editor. 2001. The dangerous toll of diabetes. Online: American Diabetes Association: www.diabetes.org.

Editor. 1996. Drinking milk regularly may cut risk of breast cancer. *Cancer Weekly Plus,* November 4, 22–23.

Editor. 2001. Raw material research: Tonalin® conjugated linoleic acid. *Nutraceuticals World* 4: 121.

Hayek, M. G., et al. 1999. Dietary conjugated linoleic acid influences the immune response of young and old C57BL/6NCrlBR mice. *Journal of Nutrition* 129: 32–38.

Houseknecht, K. L., et al. 1998. Dietary conjugated linoleic acid normalizes impaired glucose tolerance in the Zucker diabetic fatty fa/fa rat. *Biochemical and Biophysical Research Communications* 244: 678–682.

Kalman, D. S. 1998. Analyzing the latest natural weight loss supplements. *Muscular Development,* July, 96–98, 152–153.

Kreider, R., et al. Effects of conjugated linoleic acid (CLA) supplementation during resistance training on bone mineral content, bone mineral density, and markers of immune stress. *FASEB Journal* 4: A244.

Liu, K. L., et al. 1997. Conjugated linoleic acid modulation of phorbol ester-induced events in murine keratinocytes." *Lipids* 32: 725–730.

MacDonald, H. B. 2000. Conjugated linoleic acid and disease prevention: a review of current knowledge." *Journal of the American College of Nutrition* 19: 111S–118S.

Moya-Camarena, S. Y., et al. 1999. Species differences in the metabolism and regulation of gene expression by conjugated linoleic acid." *Nutrition Review* 57: 336–340.

Nature's Own. Conjugated linoleic acid. Internet Web address: www.naturesownusa.com/CLA.html.

Nicolosi, R. J., et al. 1997. Dietary conjugated linoleic acid reduces plasma lipoproteins and early aortic atherosclerosis in hypercholesterolemic hamsters. *Artery* 22: 266–277.

Pariza, M. W., et al. Mechanisms of action of conjugated linoleic acid; evidence and speculation. *Proceedings of the Society for Experimental Biology and Medicine* 223: 8–13.

Parodi, P. W. 1997. Cows' milk fat components as potential anticarcinogenic agents. *Journal of Nutrition* 127: 1055–1060.

Raloff, J. 1994. This fat may fight cancer several ways. *Science News* 145: 182–183.

Schechter, S. 1997. Fat intake can boost weight loss, if we are selective about our choices. *Better Nutrition,* June, 26.

Truitt, A., et al. 1999. Antiplatelet effects of conjugated linoleic acid isomers. *Biochimica et Biophysica Acta* 1438: 239–246.

Tsuboyama-Kasaoka, N. 2000. Conjugated linoleic acid supplementation reduces adipose tissue by apoptosis and develops lipodystrophy in Mice. *Diabetes Care* 49: 1534–1542.

West, D. B., et al. 1998. Effects of conjugated linoleic acid on body fat and energy metabolism in the mouse. *American Journal of Physiology* 275: R667–R672.

Chapter 10: MCT Oil: The Dieter's Fat

Calabrese, et al. 1999. A cross-over study of the effect of a single oral feeding of medium chain triglyceride oil vs. canola oil on post-ingestion plasma triglyceride levels in healthy men. *Alternative Medicine Review* 4:23–26.

Dias, V. 1990. Effects of feeding and energy balance in adult humans. *Metabolism* 39: 887–891.

Dulloo, A. G., M. Fathi, N. Mensi, and L. Girardier. 1996. Twenty-four-hour energy expenditure and urinary catecholamines of humans consuming low-to-moderate amounts of medium-chain triglycerides: a dose-response study in a human respiratory chamber. *European Journal of Clinical Nutrition* 50: 152–158.

Editor. 1993. Good nutrition can help athletes lose weight. *Better Nutrition* 55: 22–23.

Editor. 1995. MCT oil. *Environmental Nutrition* 8: 7.

Hainer, V., et al. 1994. The role of oils containing triacylglycerols and medium-chain fatty acids in the dietary treatment of obesity. The effect on resting energy expenditure and serum lipids. *Casopis Lekaru Ceskych* 133: 373–375.

Hill, J. O., J. C. Peters, D. Yang, et al. 1989. Thermogenesis in humans during overfeeding with medium-chain triglycerides. *Metabolism* 38: 641–648.

Roberson, A., and Parrillo, J. 1997. *Medium-chain triglycerides in sports.* Cincinnati, Ohio: Parrillo Performance.

Seaton, T. B., S. L. Welle, M. K. Warenko, and R. G. Campbell. 1986. Thermic effect of medium- and long-chain triglycerides in man. *American Journal of Clinical Nutrition* 44: 630–634.

Stubbs, R. J., and C. G. Harbon. 1996. Covert manipulation of the ratio of medium- to long-chain triglycerides in isoenergetically dense diets: effect on food intake in ad libitum feeding men. *International Journal of Eating Disorders* 20: 435–444.

Traul, K. A., et al. 2000. Review of toxicologic properties of medium-chain triglycerides. *Food and Chemical Toxicology* 38: 79–98.

Van Zyl, C. 1996. Effects of medium-chain triglyceride ingestion on fuel metabolism and cycling performance. *Journal of Applied Psychology* 80: 2217–2225.

Chapter 11: A Smart-Fat Strategy

American Heart Association. 2001. Heart and stroke guide. Online: www.americanheart.org.

Kris-Etherton, P. M. 2000. Polyunsaturated fatty acids in the food chain in the United States. *American Journal of Clinical Nutrition* 71: 179–188.

Reuters Health. 2000. Studies suggest Atkins diet is safe. Online: www.heartinfo.org.

Webb, D. 1999. Healthy diet: the smart fat makeover. *Prevention,* April, pp. 134–141.

Chapter 12: Good Fat Cooking

Crowley, M. B. 1998. Dressing up with oil. *Chatelaine,* October, p. 212D.

Maggie Greenwood-Robinson, Ph.D., is one of the country's top health and medical authors. She is the author of *Wrinkle-Free: Your Guide to Youthful Skin at Any Age*, *The Bone Density Test*, *Hair Savers for Women: A Complete Guide to Preventing and Treating Hair Loss*, *The Cellulite Breakthrough*, *Natural Weight Loss Miracles*, *Kava: The Ultimate Guide to Nature's Anti-Stress Herb*, and *21 Days to Better Fitness*. She is also the coauthor of nine other fitness books, including the national bestseller *Lean Bodies*, *Lean Bodies Total Fitness*, *High Performance Nutrition*, *Power Eating*, and *50 Workout Secrets*.

Her articles have appeared in *Let's Live*, *Physical Magazine*, *Great Life*, *Shape Magazine*, *Christian Single Magazine*, *Women's Sports and Fitness*, *Working Woman*, *Muscle and Fitness*, *Female Bodybuilding and Fitness*, and many other publications. She is a member of the advisory board of *Physical Magazine*. In additon, she has a doctorate in nutritional counseling and is a certified nutritional consultant.